COSMOS, GODS AND MADMEN

Cosmos, Gods and Madmen

Frameworks in the Anthropologies of Medicine

Edited by
Roland Littlewood and Rebecca Lynch

Published by

Berghahn Books

www.berghahnbooks.com

Library of Congress Cataloging-in-Publication Data

Names: Littlewood, Roland, editor. | Lynch, Rebecca, editor.
Title: Cosmos, gods and madmen : frameworks in the anthropologies of medicine / edited by Roland Littlewood and Rebecca Lynch.
Description: New York : Berghahn Books, 2016. | Includes bibliographical references and index.
Identifiers: LCCN 2015046285| ISBN 9781785331770 (hardback) | ISBN 9781785331787 (ebook)
Subjects: LCSH: Medical anthropology. | Religion and medicine. | Cosmology.
Classification: LCC GN296 .C685 2016 | DDC 306.4/61--dc23
LC record available at http://lccn.loc.gov/2015046285

British Library Cataloguing in Publication Data

A catalogue record for this book is available from the British Library

ISBN 978-1-78533-177-0 (hardback)
ISBN 978-1-78533-178-7 (ebook)

Contents

Acknowledgements

We are grateful to the editors of *Anthropology and Medicine* for the permission to republish chapter 9, which is derived, in part, from an article published on 5 November 2009 (available online at http://wwww.tandfonline.com10.1080/13648470903288872), and to the editors of *Transcultural Psychiatry* for permission to republish chapter 10, derived from an article published in vol. 48(3), pp. 318–35. Chapter 1 originally appeared in A. David Napier's book, *Making Things Better: A Workbook on Ritual, Cultural Values, and Environmental Behaviour*, and is republished by permission of Oxford University Press, USA.

Introduction

Divinity, Disease, Distress

Roland Littlewood and Rebecca Lynch

The four lectures that the anthropologist W. H. R. Rivers delivered to the Royal College of Physicians in 1915–16 were published after Rivers's death by his literary executor, G. Elliot Smith, professor of anatomy at University College London (and the leader of the anthropological school of 'diffusionism'). The resulting 1924 book has some claims to be the founding text of medical anthropology. Its title – chosen by Elliot Smith – was *Medicine, Magic and Religion*,[1] reflecting the idea that sickness and misfortune could no longer be considered in themselves as individual experiences: to make analytical sense of them the scholar had to take into account the way societies conceived of external events and human agency in their local worlds, including the natural world, along with the exploration of human action and human misfortune, causality and the influence on these of the ultrahuman world of gods, spirits and other extrahuman agents. In short, one had to comprehend the whole local cosmology to understand individual and collective misfortune in their proper context – and how societies responded through their shared and individual institutions.

The social anthropology of sickness and health has always been concerned with religious cosmologies: how societies make sense of such issues as prediction and control of misfortune and fate; the malevolence of others; the benevolence (or otherwise) of the mystical world; how human life may match some overarching ultrahuman principle; all this in terms of local understanding and explanations of the natural and ultrahuman worlds – as organized ritual (or other) practice, and as principles of social order and organization. With the increasing secularization of

Notes for this chapter begin on page 8.

contemporary Western societies, sickness and religion may seem to have drifted apart, yet the understanding of particular patterns of illness and misfortune still relies on general worldviews, whether they are those of naturalistic determinism or supernatural agency. Understandings of global warming or other disasters, AIDS and modern social pathologies still include certain understandings of chance and determinism, pathology and health as the fundamental building blocks of disparate world views. And all (universal at least) religions provide not only an account of extramundane beings and our ultimate justification, with prescriptive norms for social life, but also some account of the nature of humans and how they function. A theology tells us what a person is, how people differ, how and why they act, and are motivated, a theory of uniformity and difference, of reason and instinct. It offers schemata for agency, causality, influence and control; how appetites, emotions and cognitions arise, what they signify – in other words, a practical, everyday psychology. And in that, a cosmology appears particularly relevant for looking at what we call mental illness.

The culturally constructed nature of mental illness categories is well known and the work presented here covers a range of experiences that in Euro-American cultures are likely be classified as mental health issues. Locating these experiences within the environment in which they occur, with a focus on cosmologies, places understandings of them within a wider schema of local understandings, including religious ideas, revealing not only how these experiences are viewed immediately, but also how they might be incorporated into a wider worldview. Often linked to extrahuman experiences and concepts of the spiritual, these experiences may raise questions about, and illustrate alternative concepts of, agency, control and the self.

But why do we here group these understandings under the term 'cosmology': why not under 'ontology', 'philosophy' or 'psychology', or indeed 'religion'? There are levels at which these concepts overlap, suggesting the fundamental understandings of a culture, the ways of thought that categorize, link and conceptualize agents and actions in a wider worldview. Such a worldview may include human and nonhuman agents, beyond the world to the wider cosmos. While the notion of cosmologies contains elements of these understandings (religious understandings, psychological concepts and philosophical ideas and knowledge contributing to a cosmology), instead the term 'cosmology' tries to define a wider, less consistently elaborated set of ideas. Cosmologies are not fixed and unchanging notions but often made up of patchy concepts, polythetic fragments, sometimes contradictory, not always acted upon and not always clearly defined within the culture itself – Lévi-Strauss's 'floating signifiers'. They may change over time, differ by groups within a population and be influenced by new information and new categories of thought.

Over the last few years, a series of weekly research seminars at the Medical Anthropology Centre, University College London, have addressed these issues. A selection of the more significant papers from these seminars is presented here; they address, we believe, basic conceptualizations of the world that lie beneath our customary understandings of sickness and health: in short, our cosmologies, whether Western or other. The chapters cover a range of ethnographic areas, including modern communities expressly founded on certain notions of well-being and illness, and they examine notions of personhood, agency, uncertainty and power, among other questions. In so doing, the contributors seek to contextualize these understandings within the wider cultural understandings found in these areas, linking these concepts to the wider social fabric. The chapters illustrate how cosmological ideas emerge from, and contribute to, wider cultural life and the intertwining of medical and religious ideas. Indeed, the study of cosmologies bridges the areas of medical anthropology and the anthropology of religion, relevant to both and suggesting a resistance to the separation within the study of societies between what is the domain of 'religion' and what is the domain of 'medicine'.

In Rivers's book, he makes some attempt at an overall classificatory schema for what would become medical anthropology, but he notes this has to be forced and partial. Medicine, magic and religion are just abstract terms that refer to process, which, in different societies, may be separated out or more usually entangled together. As he had noted in his *History of Melanesian Society* (1914: 1),[2] 'It is hopeless to obtain a complete account of any department without covering the whole field.' He argues that as naturalistic explanations are generally lacking in small-scale societies, these three are particularly connected with each other. With the recent growth of the medical anthropology of Euro-American biomedicine, with its technical procedures, logic, research processes and personnel, we might downplay the suprahuman, but we are still concerned with questions of fate, accident and theodicy. Is the cosmos fair? Is it indifferent? And what place does an absconded god still play in it? Are we responsible? Many of Rivers's questions in 1924 are still pertinent: Is 'treatment' a logical reversal of 'causation'? Is the cause attributed to a transgression of local norms or does it just come from some malevolent fate, neglectful of human experience or indeed its actual enemy? Do all sicknesses have some ultimate logic or is there a local category standing for 'it just happens, that's the way things are'? How do systems of healing such as Western biomedicine or the Ayahuasca religions of Brazil pass beyond their original starting points to then become available for other societies?

In the first chapter, 'Why Animism Matters', David Napier recommends to us the value of an animist cosmology. Taking his starting point from Tylor's nineteenth-century suggestion that the travels of the self in sleep, dreams and trance constitute the origins for religion, Napier argues that this animist intersection of self, nature and universe makes for a more useful perception of ourselves in the world of disease. Criticized by monotheism and its monothetic scientific successors, a modern animist enchantment of nature would replace the self as a component of the outside world, now on equivalent terms with other animated external bodies. He argues the advantages in divination and in understanding creativity, in conceptualizing the 'basic' social unit of a given community and even for a legal system that attributes agency to 'accidents'. Criticizing the psychoanalytical shift from objects to mechanisms, Napier points to Jung, the therapist who might have (but never did) become central to a deep ecological anthropology, relaying the present over the past. At a point when 'universal' (i.e. Western) reason threatens to subsume human action and human rights, the importance of other modes of thought is surely significant.

Rodney Reynolds's chapter on the Black Christ of Portobelo (northern Panama) centres on its role in miraculous healing, that is points when the deity suspends the processes of the natural world and intervenes in favour of one of its votaries; just like Panama itself, an illness is both a crossroads and a potential passage. Beyond the immediate ethnic association with its blackness, the Black Christ here evokes a transvalorized provincial obscurity, local criminality and, symbolically, death and rebirth. We might note the association by Langston Hughes and W. E. B. Dubois of a Black Christ with Black American suffering. Various local Panamanians give us different accounts of the Portobelo statue's appearance in the healing arena, in which it has now become a centre for pilgrimage.

The cognitive and behavioural aspects of severe mental illness, along with the general incapacity of psychopharmaceuticals to cure it, do much to locally favour divine explanations. Two chapters here consider misfortune and the ultrahuman in rural Ghana. Ursula Read describes how the new Pentecostal 'prayer camps' extend and compete with traditional ancestral shrines in treating severe mental illness whilst competing locally against 'fetish priests', and against Islamic healing as well as the state psychiatric facilities. The power of each of these is confirmed by successful miracle stories, while being subject not only to empirical discrediting but to local tales of excessive moneymaking, deception and sexual licence. Yet, as Evans-Pritchard noted in Sudan in the 1930s, scepticism towards individual healers in no way undermines faith in others. The translation of the local gods into Christian demons does not affect a considerable interchange between the different systems in this pluralistic healing landscape. Medicines are eclectic and may be combined, but the relative inefficacy of modern psychiatry means that divine healing remains an essential alternative for the Kintampo population: 'As for sickness, it is spiritual. God did not create human beings to get sick. . . . [It] is due to Satan power, devil power, evil spirit which enters the person and make them behave that way.' Only with casting out of the spirit through the pastor's prayers can a full and complete recovery be achieved. Pentecostal churches are also significant in rumours of Ghana's *Sakawa* witchcraft as described by Alice Armstrong in her chapter on this recent phenomenon. Although not primarily concerned with healing physical or mental illness, she provides a useful engagement here with the debate on the 'modernity' of African witchcraft. *Sakawa* adepts are young boys who are rumoured to sleep in coffins, take the form of snakes and engage in cannibalism; their spirits enter their computers where they compel victims to release money. The presumed cults are led by pastors or fetish priests and Armstrong's analysis evokes not only modernity and Christianity but questions of Ghana's identity vis-à-vis Nigeria (with its famous '419' internet fraudsters), the problems of educated but unemployed young men and the extended cosmology of space and time that the internet now reveals.

Isabelle Lange in chapter 5 describes the work of a medical missionary ship as it visits Benin and Sierra Leone. The Christian idea of love/charity – which she glosses with the Setswana concept of *botho* – goes through the experience of both the seriously sick African and the visiting missionary. For the former, the congregation, comprised of body, patient and healers, resituates them in a meaningful nonstigmatizing community, 'creating a cycle where faith is healing, and healing becomes faith'. For the missionaries, they see themselves as more able to incorporate challenging and often disfiguring illnesses into their notion of the human, rebalancing the disorder created by illness and misfortune.

The Oregon members of InTouch eschew this sort of personalized deity, but their use of the practice of ritual genital manipulation points us to an attempted control over various types of 'substance use'. In her chapter, 'Addiction and the Duality of the Self in a North American Religio-Therapeutic Community', Ellie Reynolds argues that an 'addiction' is something pleasurable anybody could try once but the more you try it the more difficult it becomes to stop, for the substance takes over your own agency; following E. K. Sedgwick she argues that the substances and actions that become 'addictive' are those that remain as free choice in postmodern consumerism – but behind volition lurks compulsion. In spite of one of her informants apparently substituting her eating disorder with an addiction to sex, Reynolds argues that the central rite of manual sexual stimulation cannot really become an addiction because it is carried out with a sense of increased control and as an interhuman connection.

David Orr's chapter suggests that the Evangelical Protestant congregations of Peru would seem to offer, compared to the Roman Catholic Church, a particular appeal based on divine healing as well as salvation. And thus a reason for individual conversion might seem to lie in the instrumental seeking of health or at least the cessation of sickness. However, the local Protestant pastors do not seem to go out of their way to advertise their churches for their healing, instead emphasizing their doctrine of spiritual salvation[3] or conformity to God's way on earth. Yet the local people who gravitate to these churches tell Orr that a powerful reason for doing so is their success in dealing with alcoholism or other ills. He suggests that we find the search for both this-worldly healing and other-worldly salvation in a deeper cosmology of affliction and theodicy that exists instantiated in the daily life of the people concerned, articulated in this case through such idioms as soul loss.

What of our cosmology when God has gone? As Lynch demonstrates in her chapter, 'Cosmologies of Fear', postreligious modes of thought in the industrialized West have left us with a pervasive 'low intensity fear', which is translated into subjective anxiety: anxiety about economics, and personal debt, anxiety about our children's welfare and about the death of those close to us, about our own death and possible sickness, about the environment and the threat of war. Not so very new then, yet we now seek solace not from the priesthood but from the medical experts who patrol our risk management. The British Medical Journal has recently declared there are no 'accidents': Lynch shows us how a 'cosmology' can be approached not only through a traditional anthropology of social interpretation but additionally through psychiatric clinical data and the theories of academic biomedicine and psychology.

The final chapters nudge us perhaps to a more naturalistic perspective. Roland Littlewood's 'Functionalists and Zombis' argues, against some anthropologists, that empirical evidence for the Haitian *zombi* as a resurrected corpse is rather poor, and that those identified as such by the local population are the wandering chronically mentally ill. Yet local interpretation is that there are differences between the state of 'zombification' and that of severe mental illness. Why the apparent distinction between them? And there are problems with the classic anthropological interpretation of zombi sorcery accusations as preserving immediate social relations (the so-called homeostatic theory). This chapter suggests that both the mentally ill and the identified zombi share a relative absence of agency, marked in the local psychology/cosmology by the sorcerer's removal and retention of the *ti-bon anj*, the principle of agency, awareness and memory, as opposed to its animating principle (*gwo-bon anj*), which is retained by the hapless individual. The whole zombi complex represents the national history of Haiti, the Black Republic of former slaves who have continued to face the ever-present threat of political dependency, external intervention and the loss of self-determination. And in the necessary preservation of that historical memory, zombification is indeed 'functional'.

If the zombi chapter contains some evolutionary speculations on the origins of medical practice, Simon Dein's chapter, 'Religion and Psychosis', is a fully fledged attempt to argue that religious cosmologies and madness have identical prehistorical antecedents – the two certainly have often been conflated by materialist critics. Reviewing the different explanations that have been offered, Dein settles for agent hyper-identification: the human tendency, favoured in evolution, to 'over-attribute' agency to external powers or entities. Of obvious utility in human survival, it manifests in psychosis by the attribution of mental functioning to imagined beings outside the person. It is a major component of religious cosmologies for exactly the same reason. Like the agency-less zombi and the Peruvian with soul loss, it provides us with a model for mental illness: the madman lacks agency.[4] Similarly, an overextended theory of mind (attribution of mental states such as cognition, intention and agency to others) is shared by both religious cognition and psychotic cognition. And in both, we return to the inherent human capacities for both animism and creativity (Napier, first chapter).

The majority of the essays in this volume are ethnographic accounts revealing emic categories and the complexities of cosmologies in action. They present encounters between traditional concepts and biomedicine as well as other aspects of modernity. Some emphasize the power of biomedicine, presenting local data that might lead to questioning of biomedical categories and treatment (e.g. Lynch, Napier, Orr, Read), or even the very categories we use in the anthropology of medicine (e.g. Dein, Littlewood, Napier). Through these chapters we see differing categorizations and conflicts that occur as people seek to make sense of suffering and their experiences. Cosmologies, whether incorporating the divine or else as purely secular, lead us to interpret human action and the human constitution, its ills and its healing, in particular ways that determine and limit our very possibilities. Understanding of sickness and health, and our attempts to affirm the latter, are incomprehensible without some deeper knowledge of social being and capacity, of local causality and action, of the natural and the ultrahuman. With these essays, however partial, we hope go some way to opening up some of the multifarious associations between the world of being and the world of sickness.

Roland Littlewood is professor of anthropology and psychiatry at University College London. He is a former president of the Royal Anthropology Institute and has undertaken fieldwork in Trinidad, Haiti, Lebanon, Italy and Albania, and has published eight books and around two hundred papers.

Rebecca Lynch is a research fellow in medical anthropology at the London School of Hygiene and Tropical Medicine (LSHTM). Her work crosses the intersection of medicine and religion and she is particularly interested in understandings of morality, health and the body and cultural and scientific constructions of these. She has conducted fieldwork in Trinidad and the United Kingdom.

Notes

1. W. H. R Rivers, *Medicine, Magic and Religion* (London: Kegan Paul, Trench, Trübner and Co., 1924). Routledge has recently reissued it.
2. Rivers, *The History of Melanesian Society* (Cambridge: Cambridge University Press, 1914).
3. Salvation and healing of course have a common etymological origin in the Latin *salvus*. Christian churches have commonly emphasized purely spiritual 'healing' but also have engaged in practical this-worldly therapeutics (medieval hospital foundations, nursing orders of nuns, Seventh Day Adventists, Salvation Army) – as has Islam. Compare Greek *sotēr* (protect or saviour): soteriology, Ptolemy Soter, the Lord Protector, the Committee of Public Safety.
4. R. Littlewood, *Religion, Agency, Restitution* (Oxford: Oxford University Press, 2000); Littlewood, 'Limits to Agency in Psychopathology', *Anthropology and Medicine* 14 (2007): 95–114.

Chapter 1

Why Animism Matters

A. David Napier

Everyone who has seen visions while light-headed in fever,
everyone who has ever dreamt a dream, has seen the
phantoms of objects as well as of persons.
– E. B. Tylor, *Primitive Culture*

In 1871 Sir Edward Burnett Tylor published *Primitive Culture*, at the time a radical book for arguing that animism represented a form of religious practice in which direct evidence of sensory experience got translated into embodied rules, norms and prohibitions about the body and its local moral world. Tylor's evolutionary view had animism constituting a minimal definition of religion in which the belief that the soul travelled in sleep was evidence for religion's spiritual universality.

That the body was, as Marilyn Strathern would later have it (1988), 'partible' – that is, divisible by and through experiencing the 'self' as a part of one's behavioural world – raised developmental dilemmas for those who first engaged in studying it intellectually. This focusing on the effects of cosmology on notions of the body – and specifically on the permeability of the human body-image and body-image boundary – is why animism was then, as indeed it is now, so often thought of as an indicator of underdeveloped (or, as it were, 'primitive') inclinations.

In Tylor's terms, religion evolved from less to more abstract thinking – from multiple spiritual forces to singular, omniscient ones. Just as in psychoanalysis, the baser forms of development (what Freud defined as 'regressive' behaviours) involved a looser body-image boundary, a problem of self-knowledge and a sense of not knowing that required transcendence for advancement and growth.

But Tylor's views on animism remain, I would argue, as radical in our age as they were in his. In chapter 11 of *Primitive Culture*, Tylor lays out how animism – as a base belief that one's local environment is moved by extranormal (i.e. 'spiritual') forces – constitutes a minimal definition of religion. This definition is predicated on the notion that a soul (or another spiritual entity) can be capable of an existence independent of that physical body one perceives of as 'self', and that such an entity (here Tylor followed Auguste Comte) constituted not only humanity's primary mental state, but a condition of 'pure fetishism, constantly characterized by the free and direct exercise of our primitive tendency to conceive all external bodies soever, natural or artificial, as animated by a life essentially analogous to our own, with mere differences in intensity' (1913 [1871]: vol. 1, 477–78).

Though Tylor was later much criticized for his evolutionary perspective (where animism evolved into polydaemonism, polytheism and monotheism), his basic argument had great merit – this being that the notion of the human soul was derived from the need to explain the, as it were, 'real' experience of dreaming, and that ritual activity (most dramatically, sacrifice) derived from the twin beliefs that the soul could exist outside of the body, and that it could thus function as a free medium consorting in the space between the living and the dead.

Animism, according to Tylor, depended upon an immediate contagious connection between humans and their physical environments, as well as upon the idea that the soul could be released through sleep, trance and other liminal psychic states (e.g. hallucinations). Animism was, in other words, unlike our culturally prejudiced (Judeo-Christian, Cartesian, Salvationist) versions of personhood, highly empirical. To wit (following Williams):

> The Fiji people can show you a sort of natural well, or deep hole in the ground, at one of their islands, across the bottom of which runs a stream of water, in which you may clearly perceive the souls of men and women, beasts and plants, of stocks and stones, canoes and houses, and of all the broken utensils of this frail world, swimming, or rather tumbling along one over the other pell-mell into the regions of immortality. (Ibid.: vol. 1, 480)

For Tylor, it is as if the Cartesian artifice of a solipsistic, indulgent, individualism subverts the elegance and intellectual beauty of existential life. 'Animism' allows for the facing of experiential 'reality' head on – perhaps even in a way that today is only at all accessible to the so-called psychotic soul, as R. D. Laing had it. Viewed this way, animism requires a deep reshaping of psychology and philosophy in a manner reminiscent of the early phenomenology of Edmund Husserl. In Husserl's initial attempts at hermeneutic transcendence, a basic empiricism is galvanized by a desire, as it were, to get back to the things themselves. Husserl would not have taken kindly to what anthropologists today call 'lived experience'; he was not, like so much of contemporary anthropology, interested in how a perceived attention to high thoughts, heroic gestures and glamorous rituals could be challenged by a new and novel anthropological appropriation of the mundane and everyday.

For Husserl, assessing phenomena was not to engage in phenomenology – as it regrettably (and quite mistakenly) is for ethnography. Once Husserl realized (as Susan Sontag much later did for metaphor) the folly of pure empiricism, he reversed his focus, recognizing that meaning was only meaningful when hermeneutically embedded. This is why phenomenology, and especially existential phenomenology, soon found its home in literature and the arts, and why reflexivity – as a theoretical practice – emerged not in ethnography but in French literary criticism. As Heidegger would later claim, the words of poets constitute the house of language, and philosophers are the guardians of that house.

Indeed, those who have tried to strip meaning of its cultural content have failed repeatedly, whether they be ethnographers focused on objectivity, postmodernists focused on individual experience or cultural theorists (such as Sontag) focused on metaphorical stigma. Time and again we see that deep meaning is carried precisely in the collective social exchanges that fuel what Mark Johnson has called the 'nonpropositional and figurative structure of embodied imagination' (1987: xxxv [see also Napier 2003, 2004]). Acknowledging this truth points if anywhere to the need to see poetry, art and ritual as the precise places where any social group will invest itself most thoroughly. But if you have ever had the desire to host a nice party, you will know why ritual matters.

In refining the opportunity to make meaning socially, we allow for the possibility of focused social meaning and also new growth. As David Hecht and Maliqalim Simone so aptly put it in critiquing development, 'In their efforts to help, development workers forget, at times, that for the impoverished to survive from day to day they must ward off the hopeless reality that objectivity is constantly trying to impose upon them' (1994: 16–17). Citing an imam from Mali, they add: 'Change must discover unexpected reasons for its existence; it too, must be surprised at what it brings about. Only in the tension between the old and the new does the elaboration of a moral practice occur' (ibid.: 17). And it is, of course, in art, in music, in dance and in ritual – that is in so many non-narrative places – where those nonpropositional structures of embodied imagination lead to the invention of new moral practices. In such nonpropositional spaces, we discover socially and collectively new ways of reframing the hegemony of internationalized categories of meaning – be they diagnostic, humanitarian or just plain colonial.

What Tylor sensed already more than a century ago was that the possibilities of what could be imagined by humankind could be profoundly evidenced in the extraordinary diversity of thought elaborated through collective agreement (i.e. through culture). He also was quite aware that the myopic nature of what his own society allowed to be defined as 'religious' indicated a near total failure to elevate the senses to the natural status they universally occupy in what we now call embodied meaning. Indeed, Tylor would have found the contemporary, politically correct tendency to homogenize all of human emotional life demeaning, pretentious and offensive. As he puts it adroitly: 'the modern vulgar who ignore or repudiate the notion of ghosts of things, while retaining the notion of ghosts of persons, have fallen into a hybrid state of opinion which has neither the logic of the savage nor of the civilized philosopher' (1913 [1871]: vol. 1, 479). Courageous words, these, in both his era (which so often found very real 'primitive' peoples to be subhuman and disposable) and in ours (which finds it impossible, or unacceptable, to imagine moral worlds other than our own righteous ones).

Remarking on the appearance of Hamlet's father as a fully armoured ghost, Tylor develops a view of spirituality that is deeply phenomenal and culturally relative. Commenting on missionary accounts in North America, he offers multiple ethnographic (nearly universal) cases that exhibit how 'souls are, as it were, the shadows and the animated images of the body, and it is by a consequence of this principle that [certain peoples] believe everything to be animated in the universe' (ibid.).

In highly animistic settings, such immediate forms of connectedness – where all objects may be inhabited by sentient forces – allow humans to have an impact on the cosmos through ritual. And ritual, in turn, makes it possible to orchestrate one's local world symbolically, to develop concrete connections with one's immediate environment, to orchestrate innovative responses to moments of uncertainty and, in short, to manage one's daily world creatively. At once, in other words, Tylor dispenses with what would one day become the 'crisis' of postmodernity by taking seriously what Marcel Mauss would later call 'the techniques of the body' ('Les Techniques du corps' [1935]). Animism was for Tylor a concrete, empirical technique for relating bodily sensations to embodied meaning, embodied meaning to moral judgment, moral judgment to decision-making and decision-making to action.

Indeed, combining animism with a local cosmography allowed, as Tylor demonstrates through countless examples, both for the concrete and direct orchestration of specific responses to uncertainty, and, when focused on ritual, for the mechanical practices that made it possible to read local environmental and experiential signs for paranormal content. If, for instance, an animist applies beliefs about the auspicious or inauspicious nature of the cardinal points or to various locations in the landscape, he or she has in place an operable mechanism for responding to, and possibly warding off, disaster.

Likewise, good intentions (the 'techniques' of ritual and prayer) can compound positive outcomes by intensifying auspicious forms of embodiment. Balinese, for instance, can ward off ill health by appeasing minor spirits in the landscape, as much as avoid major disaster through orienting homes and temples in auspicious directions from which feelings and the life force that carries them (*rasa*) can flow more directly to and from a relatively permeable body now tied directly to its environment (Napier 1986: chap. 6).

However, today (perhaps more than ever) we are obliged to discount what seemed remarkable to Tylor; for the stigma against envisioning genuinely different notions of personhood is, if anything, exacerbated by global narratives about the uniformity – the universality – of psychological experience. Placing the self literally in other things and people both reshapes concepts of responsibility, and also implies that unique notions of the body might give way to culture-specific forms of emotional life – all well and good when such differences are not thought of in any hierarchical manner – very dangerous, however, when they are; for once one acknowledges that a particular mode of culturally constructed selfhood could create unique skill sets, the idea that one culture might be better at one thing or another leads to a kind of biological determinism that suppresses the uniqueness of local experience in favor of broad structural generalization. Suddenly, a fascination with creative difference is condemned by recasting it as a glamorizing of the exotic. In so doing, the experiential baby, fragile though it may be, gets abruptly thrown out as the ethnographic rabbit runs for the bolthole.

Yet, we do know in fact that body techniques can be learned and cultivated in exceedingly diverse ways, per force developing unique forms of perception and unusual skill sets. It is what we call cultural learning; those who are good at it recruit allies and succeed; those who are less good fall to the wayside, are thrown out by the pack, or choose isolation, if they can find it. Such ordering creates preferences and hierarchies ('we' become better or worse at something than 'they' are) – the very stuff of Social Darwinism (Napier 2004: 20–22).

Foucault's early view of personhood – in which the discursive practices of collectively oppressed people were subjected to the regimes of power that would have us all participating in our own subjugation – is one form of such universalizing about personhood and collective meaning across cultural domains. Though this draconian construction was revised in

Foucault's later work on sexuality and desire (for after all what is animism if not an expression of sympathy and desire?), the idea that the character of individual identity is subject to regulation by the state, by culture or by one's own compliant self was developed by many authors whose work relies on an assumed uniformity that defined persons as political – rather than, say, existential – actors.

The humanistic assumptions about the need to cultivate some global uniformity around what constitutes a person (as we see not only in human rights discourse but among anthropological and other high theorists [e.g. Butler 1991]) may build on anthropology's diverse representations of what can constitute meaning for a person, while still assuming a more monolithic and normative view of societies as producing categories to which more or less uniform individuals respond and gain or lose membership.

Though there is a long lineage (from Socrates to Foucault) of theorizing about how people react to nations and states as hegemonic managers of what can be called meaningful, such universalizing concepts of what makes a person perhaps better integrate into contemporary normalizing views of the person than into earlier ones that focused on how diverse modes of thought have concrete implications for experience, for perceptions of experience, for forms of embodiment and for understanding of the body itself (see e.g. Leenhardt 1942). Indeed, globalization (if it exists at all) may be found precisely in that form of academic historiography where the uniqueness and diversity of local forms of meaning – how they fascinated, but mostly scared the hell out of, colonizers – is erased by the mad desire to write in homogeneity across the entire human landscape.

Not only, however, do such uniform constructs clash with the desire to see the body as culturally, rather than biologically, made; they also run against many widely described institutions that refuse to be so monotonous. While the universalists may discount other views of personhood and the body as 'those exotic modes of thought made sensational by Victorian anthropology', the common Indian (Hindu) system of dual inequality, for instance (where women do for men what men cannot do for themselves, and vice versa), is much less readily dismissed; for to be a functional man or woman in such a system requires a formal relation with one's perceived opposite – a divisibility of self. Indeed, in such a system the smallest social unit is not an equally empowered man or woman, but a man and woman. Yes, replace dual inequality with plain inequality and the form of marriage becomes unacceptable; but what of the fact that the same notion, for instance, creates broad systems of care for the elderly (Cohen 1998)? What are we now to learn? And what is anthropology's (now silent) response to the global process in which the aged are accorded far more individual autonomy than they may want?

Indeed, how individual rights and duties are understood in such a context becomes particularly problematic for those who deny that culture can so deeply influence concepts of the body – and by extension of individuality and freedom of choice. However, unless one can imagine such 'partibility' – or, similarly, imagine the Maussian notion that Melanesian gift-giving (1924) involves accepting the gift as a literal part of the giver – there is little chance of seeing why a concept like animism would present such powerful challenges to our assumptions about embodiment.

What is partible for a Melanesian may be connectible for an Indian; just as what is impermeable for a European may be permeable for the Balinese who, quite happily, will describe the carrying away of an illness by shifting its pathos onto an inanimate thing that can be taken from one person and pressed upon another for good or ill-intent. The power of such modes of thinking – their unique significance for collectivities of people – is, then, evidenced each and every time we take the creative abilities of other modes of thought in any serious way.

Seeing the importance of ritual – as the dramatic and creative coalescence of forms of heightened social agreement – was not only what allowed anthropology in Tylor's day to rise as the core field in which alternate modes of thought became a real subject of investigation; it also made possible the birth of medical anthropology, through the work of W. H. R. Rivers, as the study of both human pathology and well-being; and it is worth, for this reason, quoting the first paragraph of Rivers's pioneering investigation in full:

> Medicine, magic, and religion are abstract terms, each of which connotes a large group of social processes, processes by means of which mankind has come to regulate his behaviour towards the world around him. Among ourselves these three groups of process are more or less sharply marked off from one another. One has gone altogether into the background of our social life, while the other two form distinct social categories widely different from one another, and having few elements in common. If we survey mankind widely this distinction and separation do not exist. There are many peoples among whom the three sets of social process are so closely inter-related that the disentanglement of each from the rest is difficult or impossible; while there are yet other peoples among whom the social processes to which we give the name of Medicine can hardly be said to exist, so closely is man's attitude towards disease identical with that which he adopts towards other classes of natural phenomena. (2001 [1924]: 1)

The extraordinary power, diversity and elegance of what modes of thought humans have come to conjure cannot to this day (and despite our academic erasures) be underestimated – for they are also evidenced each and every time we acknowledge the diversity of ways in which people conjure together to feel better or worse; and we gain little by simply calling such views regressive or antique. Far better, in fact, to accept their remarkable diversity and durability as evidence of a human capacity to produce forms of meaning that are as powerful as they are special. To know something of them, then, is not a luxury, but a necessity of modern life; for, regrettably, the discounting of epistemological diversity is promoted by human rights activists and human rights abusers alike.

To claim, then, that the sympathetic animation of one's environment is a 'primitive' (rather than another good) way of seeing the world is not only to reject as unknowable that which we may not already know, but to assume that it is somehow unnatural to us. Yet, as Nicholas Humphrey shows in his wonderful work on mediaeval law, the legal prosecution in mediaeval courts not only of animals and insects, but of inanimate objects, was not only done out of fearing the loss of control over the world; it was also a way of institutionalizing empathy:

> What the Greeks and mediaeval Europeans had in common was a deep fear of lawlessness: not so much fear of laws being contravened, as the much worse fear that the world they lived in might not be a lawful place at all. A statue fell on a man out of the blue, a pig killed a baby while a mother was at Mass, swarms of locusts appeared from nowhere and devastated crops. At first sight such misfortunes must have seemed to have no rhyme or reason to them. (2002: 250–51)

Now subjecting a block of wood to a criminal investigation (because it fell on, and killed, someone) seems, Humphrey points out, not at all odd from the level of believing that objects and the powers they embody not only are subjected to the same laws as those of the human world, but that the human world itself is moved by the same divine order that moves objects and other foreign bodies. Any so-called atomic view of the world, in other words, allows us to see both that we are all swayed by the same energies that are in and of everything, and that these energies can be transferred in greater or lesser degrees through sympathetic means.

To wit: any theory of unity demands that we believe that the world around us is both connected and subjected to the same rules as we ourselves are; for otherwise 'the natural universe, lawful as it may in fact have always been, was never in all respects self-evidently lawful. And people's need to believe that it was so, their faith in determinism, that everything was *not* permitted, that the centre *did* hold, had to be continually confirmed by the success of their attempts at explanation' (ibid.: 251). The stronger those connections, the more self-evident they become:

> So the law courts, on behalf of society, took matters into their own hands. Just as today, when things go unexplained, we expect the institutions of science to put facts on trial, I'd suggest the whole purpose of legal actions was to establish cognitive control. In other words, the job of the courts was to domesticate chaos, to impose order on a world of accidents – and specifically to make sense of certain seemingly inexplicable events by *redefining them as crimes* (ibid., 251–52).

As Durkheim would later argue (1912), animistic taboos that rely on sympathetic forms of magic to produce affective outcomes can have the same preventive consequences on human behaviour as the presumably more advanced forms of medical explanations in which the inexplicable is less defined as a crime than as a clinical condition. In illness, that is, the threat of a dire consequence can function equally as a behavioural proscription and as an indicator of chronic dysfunction, particularly when sociopathology evolves imperceptibly into physical or psychological morbidity.

Deviance and medicalization in this view serve the same purpose as once did the animist's sympathetic practices and taboos, even if once medicalized we treat the deviant (sociopath) and the patient (psychopath) in wholly different ways: the practice of acknowledging abnormalities while accepting that we are all connected by whatever animates us became the basis for all systems of law (Maine 1861); for laws are both what keep us in line and what provide us with good reasons for action when chaos invades life.

But I would go further: if the apparently absurd legal actions that Humphreys describes were designed to gain cognitive control over chaos, they also needed to predicate such control on the idea that all of nature shares and traffics in some same controllable thing – i.e. that what is 'of man' (as we used to say) is also of a pig, or a swarm of locusts, or a block of wood – and that it is controllable because it is shared and traded among us all. It constitutes as well as connects us.

So, being 'partible' or 'permeable' in certain cultural settings enhances our capacity to create meaningful connections (and sometimes formal obligations) locally, by reinforcing the belief that we are all subjected to the same natural energies and the laws that we create to control chaos formally. In addition, such interdependent and fluid ways of connecting to a local world can help to build strong forms of environmental responsibility – to orchestrate symbolically and sympathetically, to develop concrete, immediate and innovative responses to moments of uncertainty and to apply them to the preservation of what we value locally. In animistic settings, fluid body-image boundaries allow us to be creative in visceral ways that are less Cartesian; whereas in more Cartesian settings such fluidity is seen as not only regressive, but dangerous.

Indeed, the literature on creativity makes it clear that unlikely combinations and superimpositions of things remain central to the entire process of invention (Rothenberg 1988). New ideas derive primarily through creative superimposition of mental images (literally of neural templates that are stimulated visually), and creative thoughts, in turn, are amplified by thinking while doing – i.e. through physical activity (walking, dancing, hiking, etc.). Creativity, at least considered experientially, is an inventive, connective process enhanced by physical engagement, to the degree that its experimental domain may be dampened by Cartesian dualism, or rechanneled into religious experiences that discount social partibility in favour of charismatic practices where an animated imagination gets channeled into spirituality (Csordas 1994). It's what Tylor meant by 'monotheism'.

In the hands of the skillful animist, physically engaging powerful symbols in local space (as, for example, through conjuring) builds meaning out of empathic responses to visual and tactile stimuli; in the case of those most gifted at such forms of engagement, producing what we have come to call divination. For divination is nothing if not the creative integration in space and time of powerful structures, empirically superimposed on place and space.

Divination is, in this regard, creative: an attempt to read natural signs – to ascertain a cosmic condition through the manipulation and novel interpretation of powerful things that are manually reorganized in the interest of achieving some new understanding. Divination, in other words, is a form of creative interpretation that is nourished and enhanced through empathy and contagion. Divination promotes the condensation of power objects in the interest of forming new meaning, for seeing what has not yet been seen, for orchestrating chaos in the interest of discovery. Indeed, most highly creative people can be thought of, in one sense at least, as diviners.

Conversely, I would argue (e.g., 2004) that physical distance from domains of concrete meaning creates its own psychic dysfunction, and that the attenuation of transformations in contemporary life – the distancing that makes space for narrative rationalizing – leads not to the speeding up of transformative processes (the assumed chaotic shifts of modernity), but to a (counterintuitive) slowing down of creative growth. An attenuated separation between visceral moments of embodiment (the immediate catalysts for transformation) and their embodied resolution sometimes years later gives way to long periods of liminality that we fill with reflexive forms of interpretation, narrativity, psychotherapy, etc.

Increasingly, that is, we find (in educational, career and therapeutic processes) years, and even decades, elapsing between the immediate visceral experiences – the immediate, generative and catalytic moments that are at the core of animistic practices – and the completion of the transformations, the rites of passage, if you will, that such catalytic destabilizations generate.

As transformational structure, one might therefore posit that the kind of cosmic reading made possible by animism – where the local environment is symbolically interpreted and acted upon – is progressively abandoned in monotheistic abstraction in which attenuated change produces long periods of liminality reflected in other-worldly forms of meaning. What might have been a direct, viscerally continuous rite of passage for the animist is replaced by long periods of uncertainty. The child who after witnessing an accident desires to become a medical doctor may live out the unresolved liminality of the transformation that destabilization provoked for many years before seeing its uncertainty settled in the form of a medical school graduation, and a symbolic or real portrait of herself in a white lab coat with a stethoscope for a necklace.

Viewed this way, one can productively examine how psychoanalysis has rationalized the cultural desensitizing of the body, how it has restructured the meaning of desire and how it has discounted what we loosely call animism as an infantile and regressive behavioural mode. To recast psychoanalysis and other such reflexive processes as apologies for the symbolic and visceral disunities of modernity (rather than as curative interventions), it is useful to examine how these very practices and beliefs had their origins in the explaining away of animistic behaviours made no longer socially valid by the modernist programme. To wit: what was hysteria for Freud and Breuer, if not the initial basis for Freud's theory of psychoanalysis made meaningful by the patient's (e.g. Anna O's) inability to link through narrative the destabilizing moment (the contagious psychic 'trauma') to some conscious explanatory structure? The impulsive emotionality of the hysteric resulted, so Freud thought, from its inability to resolve itself over time. Its disconnection from life – its unresolvedness in the social world of the patient – led to its being a catalyst for psychic pathology. Find a way of making the event consciously meaningful and, it was thought, the ongoing psychic disability might be transcended.

Seen as a rationalizing response to an unresolved sense of 'being-in-the-world' (*Dasein*), to borrow an idea from Heidegger (1962), the psychoanalytic programme is not at all unlike the modern process by which traumatic victims transcend their private horrors through therapeutic writing and verbal narration (Pennebaker 2003). Interpreted from the perspective of passage rites, one could argue that psychoanalysis is modernity's facilitator – even apologist – rather than its interpreter; for pathologizing what is uncertain and liminal allows for one also to pathologize the impulsive emotionality – the hysterical reaction – that characterized and catalyzed the patient's emotional connection to a visceral moment.

For psychoanalysis, then, much of what is wrong in the world of maladjusted body-image boundary stems from the fetish, and especially from the individual's inability to relinquish it. Those of us who have willingly or unwillingly allowed ourselves to be so swayed by objects in our environment are, in a psychoanalytic sense, suffering from an inability to separate in a healthy manner the self from its behavioural world when one could as easily argue that such a separation created the need for psychoanalysis in the first place.

To engage in animistic practices has become, that is, a regressive and unhealthy pattern common only to children, so-called primitive peoples, and the mentally disturbed. And those, in turn, who actively cultivate environmental relations in which they perceive their own body-image boundaries to be fluid, are not only, the psychoanalyst would argue, engaging in regressive behaviour that must be outgrown, but risking through repetition the development of deep and damaging neurotic disturbances or psychotic episodes.

As Paul Antze points out in his essay, 'Illness as Irony in Psychoanalysis', Freud's early thoughts on illness treatment center on the process of storytelling. Storytelling builds meaning through constructing narratives of agency and its loss that

> follow the classic pattern for dramatic irony, in which the protagonist is shown to suffer from a kind of ignorance, a failure to see her real situation. And yet, as Freud repeatedly observes, there is something willful about this ignorance: 'The hysterical patient's "not knowing" was in fact a "not wanting to know" – a not wanting which might be to a greater or lesser extent conscious' (Freud and Breuer 1974 [1895]: 353). The initial splitting that gives rise to symptoms is, he insists, 'a deliberate and intentional one', usually driven by the wish to avoid a harsh truth or a difficult decision. 'Thus, the mechanism which produces hysteria represents on the one hand an act of moral cowardice and on the other a defensive measure which is at the disposal of the ego' (ibid.: 188). (Lambek and Antze 2003: 112)

But Freud's early awareness that his psychic excavations read more like short stories might have led him down an entirely different road had he been positioned to develop an interest in Mauss and van Gennep. The 'malingering' and psychosexual 'obsessive compulsions' that Freud came to associate both with hysteria and with adolescence leave little space for the deep commitment, the eternal connection if you will, of the animist with his or her social and physical environment; for a space so completely ruled by empathic desire and local responsiveness has no place in the modern world of psychic attenuation in which transformations are rarely immediate and both bodily and psychic growth are considered healthy only if one can always identify a body, a 'self', that is prior, persistent and psychically consistent.

What is more vexing still, the space between local practices and universal diagnostic categories creates a logarithmic growth in possible new disorders in which all variation can be pathologized as dysfunctional aberrations of psychiatric norms. The animist – now removed from those carefully monitored spaces of empathic exchange – experiences a pain no less deep than the indigenous person removed from his or her dreamscape, or place of ancestral connection, or shrine of animated fetishes. The success of psychoanalysis – as a widely adopted explanatory model, rather than a frequently unsuccessful mode of curing – may as easily, then, be read as a testament to the true outcome of disjoining the visceral immediacy of the animist's life from its object of meaning, as it may be read as an explanatory model having the status of scientific fact.

But where Freud placed his bet on the ability of his system to cure through narrative excision the now-dysfunctional animistic transformation, Jung perceived the more creative and positive dimensions of projection. What after all is the construct of the 'collective unconscious' other than a statement about how connecting with shared categories of thought over time and with others makes possible a retrieving and creative reshaping of a disturbing personal moment within a shared and not fully known collective space in which a troubled person can refashion himself? As the great Baroque sculptor Bernini once said of art, the greatest works of creative expression are not realized through completely independent thought, but through making something beautiful out of a flaw.

In fact, the greatest works of art, he felt, were those that addressed a flaw (a pathology if you will) in such a manner so as not only to make it the centre of the new artwork, but to make it so central that the work could not have existed without it (Napier 1992). Indeed, much of the Counter-Reformation in which Bernini was so active was built in greater or lesser degrees on the belief that lavish works of art and architecture were necessary for the resurrection, revivification and re-creation of primordial objects in yet more empathically compelling form. The job of Counter-Reformational thinkers, in other words, was not to reject the past, or what was different, but to show how its imperfections could be yet more perfectly constituted and interacted with physically – to reshape the flaws of what existed in a former time, or that exist in another tradition, into something more beautiful and remarkable still – responding to Martin Luther with the charge that the role of humankind was to take God's work and reanimate it through art.

But the higher you fly, the harder you fall. Tap into these psychic depths of human experience and you may also find yourself 'chosen' to do for others, as every shaman knows, what neither others nor their societies can do for themselves. Jung's fascination with the Book of Job and with the alchemical *coniunctio* are vivid examples of how those who have chosen – i.e. been chosen by destiny – to reframe humanity's relationship to its experiential world may find themselves searching for a Holy Grail, especially that Grail upon which was, according to von Eschenbach's twelfth-century *Parzivaal,* inscribed the charge that 'any Templar chosen by god's hand to rule over a foreign folk must refuse the asking of his name and help them to their rights'.

Choosing or being chosen – that is, to rethink what might be 'animated' by completing such deeply religious tasks – carried with it not only the possibility of creating something very new, but of risking destruction and failure. Note, as did Mauss, with what trepidation and feelings of nausea and ill health a Trobriander must accept a powerful *kula* object from a trading partner. Note in particular the requirement that he make good on the exchange by bringing yet more fame to the animated object he receives, and one senses immediately the burden of creativity – that is, why those who reanimate the things we collectively share may suffer deeply in so doing. This may well be what Nietzsche had in mind when he described humankind as being doomed to create.

To conclude: if narrativity and therapeutic writing are viewed as curative processes (if not our cultural apologetics) in which a body seeks out its own continuity through storytelling, what can we learn from the desirous commitments of the animistic, regressive body? Are there places in modernity where Tylor's challenge and charge that we think beyond culture can be realized – where we learn from others new ways of taking responsibility for the local worlds we inhabit? Does, in other words, the acknowledgement of the creative possibilities of animism – its concrete superimpositions of the past and the present through ritual focus – offer us any insight into what it is we seek to recover and reshape when we look longingly at traditional systems of knowledge, or desire ourselves to reconnect with our local environments in ways that are more than cerebral? Is it only the awareness of what we have lost that we have to gain by reconsidering what animism makes possible?

Have we not redoubled the difficulties of recapturing and recreating something new when we demand of ourselves that we perceive the future, as did ancient Greeks, as a pushing forward of the past into something both new and eternal? How can such a daunting task be accomplished when it also requires some self-conscious technique for literally losing one's consciousness in the things of this world?

How do we, furthermore, come to grips with the depth of pain caused when an animist's local symbolic world is disrupted, when the Australian or Native American is forced to relocate in the interest of one or another development scheme? How can awareness be reconstituted when, as a homeless friend of mine once put it, 'you do not even know what you do not know?'

These are, it seems to me, important and intractable questions the further elaboration of which must, alas, be left for another time and place.

A. David Napier is professor of medical anthropology at University College London, director of the University's Centre for Applied Global Citizenship and director of its Science, Medicine and Society Network. His special interests in applied research include assessing vulnerability, primary healthcare delivery, human well-being, caring for ethnically diverse populations, migration and trafficking, homelessness, new and emerging technologies, immunology and creativity in scientific practice. He has published on law and anthropology and intellectual property and biodiversity, and is the author of six scholarly books and numerous book chapters.

References

Butler, J. 1989. 'Foucault and the Paradox of Bodily Inscriptions.' *Journal of Philosophy* 86(11): 601–7.
Cohen, L. 1998. *No Aging in India: Alzheimer's, the Bad Family, and Other Modern Things*. Berkeley: University of California Press.
Csordas, T. 1994. *The Sacred Self: A Cultural Phenomenology of Charismatic Healing*. Berkeley: University of California Press.
Durkheim, É. 2008 [1912]. *The Elementary Forms of Religious Life*. Oxford: Oxford Paperbacks.
Freud, S., and J. Breuer. 1974 [1895]. *Studies on Hysteria*. London: Penguin.
Hecht, D., and M. Simone. 1994. *Invisible Governance: The Art of African Micropolitics*. Brooklyn: Autonomedia.
Heidegger, M. 1962. *Being and Time*. New York: SCM Press.
Humphrey, N. 2002. *The Mind Made Flesh: Frontiers of Psychology and Evolution*. Oxford: Oxford University Press.
Lambek, M., and P. Antze (eds). 2004. *Illness and Irony: On the Ambiguity of Suffering in Culture*. New York and Oxford: Berghahn Books.
Leenhardt, M. 1942. 'La Personne Mélanésienne.' In Leenhardt, *La Structure de la Personne en Mélanésie*. Milan: STUA Edizioni, pp. 92–120.

Maine, H. J. S. 1861. *Ancient Law: Its Connection to the History of Early Society*. London: John Murray.
Mauss, M. 1935. 'Les Techniques du Corps.' *Journal de Psychologie Normal et Pathologique* 32: 271–93.
———. 1967 [1924]. *The Gift*. New York: Norton.
Napier, A. D. 1986. *Masks, Transformation, and Paradox*. Berkeley: University of California Press.
———. 1992. *Foreign Bodies: Performance, Art, and Symbolic Anthropology*. Berkeley: University of California Press.
———. 2003. *The Age of Immunology: Conceiving a Future in an Alienating World*. Chicago: University of Chicago Press.
———. 2004. *The Righting of Passage: Perceptions of Change After Modernity*. Philadelphia: University of Pennsylvania Press.
Pennebaker, J. W. 2003. 'Telling Stories: the Health Benefits of Disclosure.' In J. M. Wilce Jr (ed.), *Social and Cultural Lives of Immune Systems*. London: Routledge, pp. 19–34.
Rivers, W. H. R. 2001 [1924]. *Magic, Medicine, and Religion*. London: Routledge.
Rothenberg, A. 1988. 'Creativity and the Homospatial Process: Experimental Studies.' *Psychiatric Clinics of North America* 11(3): 443–59.
Strathern, M. 1988. *The Gender of the Gift*. Berkeley: University of California Press.
Tylor, E. B. 1913 [1871]. *Primitive Culture: Researches into the Development of Mythology, Philosophy, Religion, Language, Art, and Custom*. 2 vols. London: John Murray.

Chapter 2

Spreading the Gospel of the Miracle Cure

Panama's Black Christ

Rodney J. Reynolds

This chapter considers miraculous healing performed by the Black Christ of Portobelo, who is situated in a small, Caribbean coast town in the Republic of Panama. The sculpted, wooden and lead figure is a life-sized statue of Jesus the Nazarene in a posture from one of the stations of the cross. It is installed in Portobelo's cathedral, which is named after Saint Philip and also lends the village its proper name, San Felipe de Portobelo. The figure is known locally as El Nazareno (The Nazarene) as well as the Black Christ. The Black Christ's miracles are in response to prayer and other acts of devotion undertaken by the Christ's followers.

I argue that the Black Christ functions as an emblem of the logic of healing in Panama. That logic positions sickness as an opportunity (for the Black Christ's devotees) to demonstrate that sickness may be resolved by instantiating curative therapeutic pathways that biomedical diagnoses would seem neither to offer nor to recognize. I suggest therefore that sickness posits the restoration of health as indicative of a process of worldmaking. I use 'worldmaking' in the sense long ago argued by Nelson Goodman (1978), in which the uncertainties associated with sickness counter-intuitively multiply the possibility of healthy outcomes and so worlds in which those outcomes are true.

Prayer is one act of devotion used in the context of illness and sickness that I will discuss below. Illness and sickness create the opportunity for devotional acts of prayer to bring the intended outcome of praying into being, precisely because states of sickness are vulnerable to change. In this chapter I use illness and sickness as interchangeable concepts and understand them to be a cultural resource through which the self and the world inhabited by it are both necessarily *already* marked for change and transformation. We know that illness experiences are characterized by shifting perceptions of one's state. Thus, with respect to illness generally and to chronic conditions in particular, any given person lives in constant states of 'relapse' to illness or 'return' to whatever the person's expectations of normal health (Garcia 2010) might be.

'Relapse' and 'return' are borrowed from Angela Garcia, who has worked extensively on addiction in Mexico. Her book, *The Pastoral Clinic* (2010), elaborates the analytical use of these concepts as frames within which addicts encounter themselves forever haunted by the not-unexpected possibility of rebounding into addiction from having (temporarily) overcome substance misuse. She implies in her ethnography that recovering addicts tend to experience return to 'normal' health as a deficit because the potential for relapse and its concomitant connotation of personal failure is ever-present. I apply Garcia's concept in a broader way than she intended it. I wish to emphasize that the utility of illness emerges in how it reminds us of our potential to be other than we are and that specific and sustained acts may be required to effect whatever change or maintenance we desire. With the terms 'relapse' and 'return',[1] I hope to characterize 'living' as an ongoing process of situating oneself with respect to what is possible in terms of health outcomes.

The broad quality of this argument affirms the logic of Thomas Csordas's familiar and important description of therapeutic process as operational at the 'margin of disability' because disability is 'constituted as a habitual mode of engaging the world' (1997: 71). As Csordas states, the wheelchair-bound man is able to walk because he already knew he could, doing so in response to his own motivation may have been too painful or possibly some other sanction existed against performing this act with any regularity. In other words, healing was positively demonstrated only because the desired outcome was already possible. The trick, as it were, is to motivate its occurrence.

Meyer Fortes (1987) long ago made a similar point in his work among the Tallensi in describing how one copes with a destiny one does not like. Divination, and subsequently the proper organization of ancestors, resources and invocation of ritual practice, 'sweeps away' bad predestiny thus allowing for its reversal. This assumes bad predestiny has been properly recognized, captured and materially fixed for ritual handling, while also recognizing and building a good destiny.

Returning to Csordas, where his analysis fails is in the exclusion of miraculous claims of spontaneous and total healing from his description of a therapeutic process (1997: 72). He has to have this exclusion because this kind of healing, on the one hand, would not necessarily be responsive to intervention and, on the other, such an outcome could not be already known. My reading of Fortes (1987: 169) runs up against a similar problem, since the Tallensi seek the guidance of a diviner who will enlist the goodwill of ancestors and so authorize the ancestors' necessarily effective ritual interventions on behalf of one who suffers; thus what is impossible – in this case the ancestors ignoring the petition – has already been excluded. How then does one confront the problem of the miracle, which is characterized by having overcome the limits of possibility?

The answer would seem to be, given Panamanian practice, to deny the possibility of impossibility. The 'miracle cure', an abstraction, is thus refigured as material and understood as a fundamental condition of the living process. Through miracles, the world is made plural. Goodman reminds us that worlds are made, 'not from nothing, after all, but *from other worlds.* Worldmaking as we know it always starts from worlds already on hand; the making is a remaking' (1978: 6, emphasis in original). 'Relapse' and 'return', therefore, I intend to function as shorthand to capture the sensibility of the oscillation that occurs through living only in terms of what is possible and the need to perform regular acts that will ensure the possible remains a possibility. These terms are part of a system of classification built with a sensibility of mutual implication. I understand them as opposed to notions of recovery, which would seem to connote a more singular trajectory as one state simply having overcome another.

Panama and Portobelo in History

Panama is both a crossroads and bridge. It links North America and South, the Atlantic and the Pacific. Outsiders often reconcile this duality by reference to Panama's association with its canal. Easy resolution of Panama into a conduit or passage through appeal to conceptual constructions of the Republic centred around its canal is problematic, however. This is because such representations occlude and exoticize Panamanians' understandings and experiences of the worlds within which they live their day-to-day realities.

Panama's complexity arises from its territorial characteristics that produce propinquities out of the circumstances of the distances and differences found in the world beyond the Republic's borders. The fact that Panama provides a means for resolving the difficulty of how east becomes west (via the trope/reality of the canal, for instance) should not be allowed to obscure the ways in which Panamanians would construct the local world(s) in which they live. This chapter engages with the problem of worldmaking, given a globalizing imperative, by focusing on illness and miracle cures as a modality that underscores how Panamanians use imagistic and material practices to reconstruct a single lived world into something more plural, much as a crossroads allows one to continue a journey via multiple routes.

Christened 'Puerto Bello' by Columbus in 1502, who had sheltered his ships from a storm in the village's natural bay, Portobelo today is little more than a village with about three hundred people. Its architecture is dominated by stone, Colonial era ruins and closely huddled, low one- and two-story, British Caribbean–style wooden houses. These are built on gentle slopes that lead to the sea on one side and up into jungle on the other. Behind the church of San Felipe, where the Black Christ is housed, the local residences become concrete and meander for about half a mile along a network of narrow, rough, hard pack, gravel and potholed paved streets.

The region to which Portobelo belongs was first identified and exploited by white Europeans, but today is associated most closely with black Caribbean cultures and people and with the descendants of African people whom the Spanish colonials had enslaved. Some of these, known in English as Maroons, freed themselves by walking into the dense tropical areas surrounding the village. There they established communities called (in Panama) 'palenques'. By emphasizing racial identifications between this territory and its historical and present populations, I want to shift attention to how those identifications transformed over time. How perceptions of the world are organized may change in relation to the symbols and physical states through which the world is known. I also wish to underscore the contested way in which Panama is linked to outside ideas and events through the Christ of Portobelo.

A Black Christ

A photocopied document I was given by a friend during my fieldwork (Montenegro, Perdomo and de Castillo 1983: 40–41) argues that the designation 'Black Christ' originates with 'gringo' (which generally means 'foreign' in Panama, but is usually directed at Americans) soldiers who arrived en masse – between six and seven hundred men – in Portobelo at the end of World War II on 21 October, the day on which the Christ of Portobelo is venerated with a village procession. The soldiers had apparently identified Portobelo with its Christ as a site at which they could perform an act of devotion to thank God for having ended the war. Montenegro, Perdomo and de Castillo's original text does not state whether or not the soldiers were devotees to the Christ of Portobelo, so broader questions of the soldiers' motivations in choosing Portobelo for their devotional act remain unknown. Nevertheless, apparently overcome with emotion at being able to honour their promise of veneration, the soldiers began to shout, 'Viva el Cristo Negro!' Since this time, the figure of the Christ has been identified with the designation 'Black Christ'.

The authors go on to assert that this custom is locally disapproved of and that devotees and people from Portobelo refer to the Christ as 'El Nazareno' or 'Jesús Nazareno', while foreigners use the more colloquial and racialized term. Though I did not find this sensibility strictly adhered to in my fieldwork, undertaken twenty years later, the logic of its assertion is generally consistent with my findings of the ways in which Panamanians distil those who belong to their local world from those who do not. 'El Nazareno', as opposed to the 'Black Christ', pulls the figure in two interpretative directions. The first associates it with a straightforward Catholic veneration practice inflected with local tradition and custom. The second links the figure of the Christ to a broader trope of the Black Christ that exists as a miracle worker and symbol of morality and equity throughout the Americas and the Philippines.

Nelson Goodman (1978) recognizes that one way of producing worlds is through the conventional organization and interpretation of symbols. Reinterpretation of a symbol then may involve the work of shifting its frame of reference, which, as Goodman points out, emerges not from the thing described but from a pre-existing system of description. The gringo soldiers invoked a racial system of description for the object of their veneration that was at odds with local practice, even as they reinforced local practices of devotion through acts of procession, holding the figure of Christ aloft and shouting words of praise (Montenegro, Perdomo and de Castillo 1983).

New worlds do not arise every time an old one is dismantled and reassembled; but distinctions between worlds do arise when everything that belongs to a world that has been undone fails to appear in another which it has begotten (Goodman 1978). By framing the Christ of Portobelo as 'Black', the soldiers brought to it from the world outside of Panama a role in a world remade by war that had not previously existed for it in which the 'Black' moniker signified something meaningful.

Simply rearranging the constituent content of a world would only produce a new version of what had been. Instantiating a 'Black Christ' tapped into a gringo racialized logic that would have been familiar in Panama and yet apparently novel when applied to devotional religious practices. The Black Christ was made through its association with a way of understanding the world that Panama as a nation actively constructed itself against.

El Nazareno's Arrival in Portobelo

There are multiple stories of how the Christ of Portobelo came to Portobelo and was installed in the church of San Felipe. There are three main accounts and several variations of each of those.

The Box and the Storm

Some say that there was a boat bound for Cartagena and that every time it tried to leave port a violent storm would commence requiring a return to port. On the fifth try the ship was on the point of sinking and the crew decided to lighten their load by throwing a big and heavy crate overboard that they were carrying as cargo. A fisherman found it floating and brought it ashore but did not open it. The town was subsequently visited with a plague of smallpox. During the plague townspeople opened the crate and saw the Christ inside. They removed the statue and venerated it and installed it in their church. The smallpox outbreak ceased immediately and gave to the Christ a reputation as a healer.

The Box and the Epidemic

A variation on the story above maintains that fishermen found the box floating in the sea during a cholera epidemic. Inside of the box was the Christ. They took him out and installed him in the church. Almost immediately, the epidemic ended and those who were sick recuperated rapidly.

The Mix-Up of Images

A third legend maintains that the Taboga church, which is located on Panama's Pacific coast, had ordered an image of Jesus of Nazareth from a source in Spain. At the same time, the church in Portobelo solicited an image of St Peter from the same artisan. The crates arrived by ship but somehow were switched and St Peter ended up in Taboga and the Nazarene in Portobelo.

Written sources date the Christ's arrival as 21 October 1658 and say that he arrived on the waves of the Atlantic washing up on the shores of Portobelo. Others say that an indigenous man, Kichimbanchi, discovered the saint floating on the water and dragged him to shore. A woman I spoke with objected to my asking for her 'version' of the origin myth of the Christ:

> I don't have a version. I know what I have heard and I think that is what the whole world knows. Fishermen found the image. There was a ship that sank before it could arrive in South America and the crate with the Christ inside floated to Portobelo. Also he became dark because there was a fire in the church. The natives of Portobelo adorn him with forest flowers and music while the people from Colón who have a Caribbean heritage decorate him with gold. I am always moved when I see the Christ. I don't know what to say or how to explain it. The force or the impact of the Christ in the people who undertake mandas, it quiets me.

I have little doubt that the Christ dates from the seventeenth century. I have not yet been able to verify outbreaks of disease during that century around the date that is given for the Christ's arrival, but such is not implausible. All of the narratives of the Christ's origin turn on the figure–ground of water and land. Each implies the other. The water, I suggest, was the means of arrival for the Christ, and until the 1980s the sea was the way in which one would usually arrive to the town because of virtually impassable roads. The sea and its currents are the 'roads' that bring the devotees. Many of them will carry replicas of the box in which the Christ is normally housed, which in turn recalls the box from which the Christ was freed through an original act of devotion. The box is a constraint or a problem that also functions as an indication of extreme locality – of the power of being in a space or situation from which one wishes to extricate oneself or another. By opening the box, a new world free of epidemic disease was brought into being; as was a reciprocal relationship of necessary symbolic and devotional acts between the town, its inhabitants and the Nazarene who had come from the sea.

Often saints are in an indexical relation with a given territory. Saints are placed or they show up when and where they like of their own volition – sometimes in response to having been called or prayed for. The index arises as a result of the fact that the saint has appeared in a given place *and* has been seen or because some authority of the Church installed the saint in that location. The principle of indexicality is not only concerned with the installation of a saint in a given place but with rendering the link between saint and place natural: 'I have seen him [the Black Christ] a couple of times and his gaze seems real to me. No one could remove that image from Colon and when they tried it nobody could do it because inexplicable things would happen.' Linette's matter of fact observation is an oblique reference to how the Black Christ came to be in Portobelo. More significantly, however, her comment addresses the fit between the Christ and place that once established cannot be broken by human intervention.

The relationship is as it is supposed to be. Any attempt to make it otherwise results in punitive supernatural intervention, as the origin tales illustrate.

El Nazareno

El Nazareno stands, wears a crown of thorns and has not yet been stripped of his purple robe as described in the Gospels. On the day of his celebration in October, he is joined on his altar by a helper who bears the weight of the cross with him. I have no information about this second figure, but must presume it to be Simon of Cyrene, historically understood to be a north African Jew who was pressed into service to help carry the cross after Christ fell.

The Simon figure wears a full beard, a hat similar to a chauffeur's cap (some years the cap is a Fez) and wears a khaki garment with a maroon sash that looks like a uniform. The priests affiliated with the church in which the Black Christ resides rotate and there is little retained information about the history or iconography of the figures that I have been able to identify to date. This second figure of 'Simon' atop the altar in the bowtie is never mentioned by anyone I spoke to about the Nazarene.

Chueng, a devotee, stared at me in surprise when I asked her about 'Simon' and peered at the figure closely as if she were seeing it for the first time. 'What strange clothes! A white guy dressed in white with a bowtie. How bizarre. He looks like he is wearing a uniform, like he is part of a unit of some kind. . . . The truth is I ignored it [the figure].' Despite the second figure's apparent invisibility, its coupling with the Nazarene would suggest that the Portobelo Christ enacts a station of the cross, linking the Christ to a specific place, conceptually. The interior of the Nazarene's cathedral has photos of the Stations of the Cross hanging on the walls and the October celebration's altar tableau of the Christ with Simon is number 8 in the cycle. 'The Nazarene is the image of the figure of Christ on the road to Calvary that comes from the gospels,' Erika, another devotee explained to me. 'Remember, those figures of the saints and the paintings are to teach with. The image of the Nazarene is the moment of [Christ in] transit marching towards the cross. Jesus alive, suffering. The crucifixion is a different image, without robes.'

Contemporary Christian belief codified in seventeenth-century Spain holds that there are fourteen stations. Biblically, they emphasize the interplay and struggle between people, light and dark, true and false. Belief in the Stations of the Cross spread from mainland Spain to its colonies and possessions from the seventeenth century onward. The Stations of the Cross were intended as sites of meditation to contemplate Christ's final journey. The station associated with Simon is linked to verses from the books of Mark and Matthew; obedience being the meditative theme.

The Christ's wooden cross lies with its weight on the figure's left shoulder, its length trailing behind. The face and visible part of the chest are made of lead and reach to the figure's mid-torso. These are attached to a vertical wooden block hidden under the Christ's robes that is never seen. The black wig covering the head falls to the shoulders and is made of human hair. The feet are hidden by the robe, but like the hands are carved from cocobolo, a rare type of tropical rosewood. These are periodically replaced by caretakers, as is the cross the Christ carries.

The right hand, palm-up and turned slightly inward, is held at the waist. The arm is bent in the gesture of cradling something that has been recently removed. The left arm is also bent at the elbow but sticks straight out in front of the figure and ends in an open upturned palm. The gesture is one of supplication. The hands and the other visible parts of the figure are the colour of a rich roux. The neck and head emerge from the hollow of a purple robe trimmed in white lace, suggesting not only an association with royal sumptuary conventions but with the Catholic order of Dominicans as well.

The public never sees the Christ without his robes and I doubt there is general awareness that the entire form is not carved. I happened to catch a glimpse when it was being moved from storage. It is installed in a heavy, carved wooden altar trimmed with gold in the western part of the apse of the whitewashed Iglesia San Felipe, the tallest structure in Portobelo. Another Christ floats above the altar suspended on the wall in the posture of crucifixion. During the year, El Nazareno is kept behind glass in a wooden box atop a rack that holds the candles of devotees. The robes the Christ wears are donated. The ones from previous years have been installed in a museum in the town that has recently opened.

The parish priest says that he is overwhelmed with offers to dress the Christ each year and often penitents will offer to arraign the figure as a devotional act. Local women separately gather together and make floral adornments for him. The priest indicates that he is excluded from this activity that is entirely locally organized. This responsibility is handed down in families, just as the task of shouldering the altar is divided among the local men. A mark of belonging is to be asked to join in the bearers of the altar on 21 October.

Struggle in the World

El Nazareno leaves the church only twice a year. One of those days is on 21 October at 8 pm, complete with the tolling of bells, drums, shouts of 'Viva!', songs of praise and blaring trumpets. On 21 and 22 October every year, the village of three hundred accommodates tens of thousands. These visitors fall into four broad categories. Some want to thank the Christ for having performed a miracle. Others ask a favour and make a promise, while another group seeks to repent something and to change their lives by indebting themselves to the Christ and his power. However, many people come to Portobelo during these two days for the spectacle and understand believers' practices as little more than a curiosity.

On 21 October the Christ is placed on a litter that is carried on a route around the town that circles the main square. The men of Portobelo hoist the litter onto their shoulders to shouts of 'Viva!' from the people, who will have packed the church and who line the processional route. The bearers sway and take three steps forward, and two steps back, porting the figure out of the church around the plaza and then back to his resting place. The procession is slow and takes many hours to complete. I have been told that the three-step-two-step shuffle is to represent the weakness of the first Portobeleños, who, suffering from disease, took the Christ out of his crate and installed him in the church of San Juan de Dios, later transferring him to the San Felipe Cathedral (Montenegro, Perdomo and de Castillo 1983).

The stagger step is also a material metaphor for the weight of the burden carried by Christ and imitates how he might have moved when shouldering the cross. For the men, their stagger is no pretence, the task they have volunteered to undertake weighs literally on their shoulders, bows their knees and presses their bodies to exhaustion. Significantly, their movements and actions mimic the drama enacted in the tableau of the litter their shoulders carry. Their act of carrying El Nazareno makes them correspond symbolically with Simon, whom they support from underneath, along with the Christ.

The biblical reading of Christ's encounter with Simon is intended to give greater richness to the associations between the cross as representative of the hardships of life lived and the pain and suffering one encounters as a result of living life. Christ's burden was lessened by Simon just as Simon's was lessened by Christ. This is explicitly a key tenet of Christianity, which is filled with language of 'giving up or sharing one's burden with Christ', 'laying down one's burden', 'not being given crosses one cannot bear with Jesus's help', etc. The intended message is clear. In reciprocal relation with Christ there is strength, especially when one walks in his footsteps. The men who carry the litter in Portobello do not walk in Christ's footsteps from behind, yet their role is still one of support and they are offered spiritual blessing and health as a consequence of their labour.

The great Puerto Rican vocalist Ismael Rivera, nicknamed 'Maleo', credited El Nazareno with enabling him to overcome heroin addiction. Rivera had been arrested in Panama for heroin possession and he served time in prison there. During this period he likely became a Nazareno devotee. Over the next sixteen years, following his release from prison, Maleo would make an annual pilgrimage to Portobelo to join local men in shouldering the burden of carrying Christ and Simon around the town (Flores 2004). The ritual act both produced and maintained the world in which Rivera struggled to live.

> After it was over, we found Maleo near the steps of the Church where barbers were shaving and cutting the men's hair. He showed us the bruises on his shoulder from the platform, and I asked him why he was cutting his hair and beard. 'I grow it all year as part of my promise to El Nazareno, and then I leave him my strength so that he can continue to help me.' (Ibid.: 73)

Miracles

Salome's family emigrated from Chile to Colón, the capital of Panama's Caribbean coast and entry to the country's canal, when she and her older brother were young children in the early 1980's. Salome and I worked together at Panama's Contemporary Art Museum during my fieldwork between 2002 and 2004 and we became friends. Having grown up in Colón, which is about thirty-five kilometres away from Portobelo and, like Colón, situated on Panama's Caribbean coast, I thought Salo might have some opinion about Portobelo's Black Christ. I hadn't known she had any link with the Christ when I initiated the subject. 'Well,' Salome began in response to my question, 'He has an expression of pain and for me it is unquestionable that he is *super* miraculous, because he saved the lives of my mother and grandfather. There are *so* many testimonials to the fact of his ability to work miracles, which is why so many people make pilgrimages there.'

The reputation of the Black Christ of Portobelo as instantiating miracles in people's lives is widespread. During the annual celebration of the Black Christ on 21 and 22 October, some fifty to seventy-five thousand people, according to media estimates, arrive annually in Portobelo for the Christ's procession. The majority of these come from Panama's provinces, but many travel to Portobelo from abroad as an act of devotion.

Before responding to my questions for elaboration about her family's experiences with the Christ, Salo shared her understanding of where El Nazareno had come from:

> They say that he arrived in a boat, during the colonial era I believe, and because there were a lot of skirmishes during this period [between the Spanish and European buccaneers and pirates] the figure had been brought to Portobelo so that it could be relocated elsewhere by ship for protection, but a storm began that was so tremendous that when the ship tried to sail away it was impossible and so the crew decided to leave the figure in Portobelo. People say that he [the Black Christ] wanted to stay in Portobelo to care for the population. Portobelo has a kind of lethargy, and maybe for that reason the place would need guidance and prayer. It is true that there is a very strong energy when you enter the church which is very, very old and well, seeing is believing, you should see it for yourself.

I told her I knew Portobelo well and had been there many times. She nodded and continued,

> The doctors told my mother that she was going to die when she was going to give birth to my brother Diego, that placenta previa, which she had had previously, would cause her death, and miraculously she was saved! All my family performed an act of devotion and prayer (*una manda*) to the Christ and the placenta previa disappeared. I remember I was four years old and we went to complete the offering (*fuimos a pagar la manda*). Everybody went. Everyone. My uncles, my granny, my father. Everyone.

Placenta previa occurs when the placenta attaches to the lower part of the uterine wall, partially or totally covering the cervix. It is usually diagnosed by bleeding during the second and third trimester of pregnancy. The condition is fairly uncommon, occurring in about one pregnancy in two hundred, but women who have previously had the condition are at greater risk for recurrence with subsequent pregnancies. Depending upon the severity of the condition, it may not prevent vaginal birth and under certain circumstances may be self-correcting. In acute situations, the baby may be taken by caesarean prior to term so as to prevent uncontrolled haemorrhage and death of the mother.

The manda is the offering given to the saint in exchange for an answered prayer. It is both an act of faith and indicative of a solemn engagement of responsibility. Patricia, a woman in her twenties who recently emigrated to Texas from Panama, commented:

> Me? No, I have never done a manda because I never had the need. The people who believe in that do so because they have grave illnesses or incurable

> illnesses or other extreme sicknesses. That is what people say. I don't have any personal experience and can't really say. I neither believe nor don't believe. But if the people say that it is true, they must have a reason, but I don't have any proof. But it is the same faith and it can move mountains. You know Panamanians are very faithful. They are believers, but maybe we are losing that generation after generation, you know.

Salome went on, shifting her focus:

> And my grandpa who was seventy-five years old was run over by a car and was given three days to die, and if he didn't die they would have to cut off his leg and again the Black Christ intervened and saved him, saved his leg and now he is perfectly fine, lucid and everything. And that time too, I went to '*pagar la manda*' with my mom. It was truly a miracle. He was a month in a coma and after that in intensive therapy.

So the Christ is like a really strong medicine in case of emergencies, I suggested. She laughed and agreed. The curative healing that she described is typical of the examples recounted to me as I moved through the throngs of devotees speaking with those who would talk to me about their reasons for being there. The miracles Salome recounted would have fallen outside of the system elaborated by Csordas and introduced briefly above, because there is no verifiable connection to a self-activated process via a healing intervention. I asked if the manda had been completed during the October celebration: 'No, no, none of that. We just prayed and afterwards went by car to the church to see him [the Black Christ] and we carried flowers, candles and photos of my grandpa. We prayed once again when we got back home. We prayed a novena, nine days we prayed.' A novena is a catholic devotional act of prayer repeated for nine days – nine carrying symbolic value in the catholic faith and representing in some traditions the period of Christ's gestation. It is also linked to mourning traditions, and novenas are usually classed in one of four ways: mourning, preparation for a rite or ritual, acknowledgement of punishment having been rescinded and some act of sin forgiven and finally novenas can serve as a vehicle for prayer and the attainment of greater understanding of how to pray and for what, this knowledge leading one necessarily closer to God.

How did you choose to carry those items, I asked: 'Well, it just occurred to me, almost always you carry those things. The candles for light, guidance and wisdom, the flowers for adorning the altar and the photo so the Christ would know very well who my grandpa is.' And why the Christ of Portobelo, I probed, and not the Christ of Atalaya, also a Nazarene and usually associated with Panama's Interior (the part of the country on the other side of the canal from Panama City and several hours distant from the capital) provinces and Spanish heritage, or, alternatively, the Christ of Esquipulas, a crucifixion figure and linked territorially with Mexico and Guatemala, but also present in Panama:

> Because we already knew that the Christ of Portobelo was a powerful miracle worker (*super milagroso*) because of the experience with my mom. Those other Christs, I am not really familiar with, in any case it's the same Christ, it is Jesus but in different places, different statues, it is the faith that people place in the figure but it is still God, just like with the Virgin, the Virgin has many names but she is always the same.

The location of the saint matters, since when a prayer is answered, attending the saint in person, being in its presence, seeing it and being seen by it, leaving a votive offering of some kind – flowers, money, incense, candles or photos – is necessary to close a circular relationship of reciprocity. To deny the Christ what it is due, which is tantamount to an attempt to contravene its will, is to invite punishment. 'You must complete the manda, that is a serious thing. Once you make a promise to God it must be kept or things will become very hard for you, much worse than they were before.'

For Salome and her family, these events of curative healing were miracles not so much because the certainty of a bad outcome (that could not be altered by medical intervention) was given; rather, the possibility of that negative outcome not coming to pass only existed *after* a world had been created in which the negative outcome did not exist.

Whether miracles occur is not in question. Neither is the means for bringing them about. The reputation of El Nazareno for enabling miracles is an ontological certainty. The mechanism, as it were, turns on the conventional organization of social and symbolic resources, such that the simple act of living produces itself as grounded in acts of reciprocal responsibility that are symbolically marked.

Operationalizing the symbolic capacity of flowers, photographs and candles, and undertaking specific acts of conduct and comportment as fulfilment of a promise, takes advantage of the knowledge that those components of everyday life can be repurposed and redirected. In doing so, a version of the world emerges that is contingent upon whether a miraculous event will occur.

Miracles are by definition not given and yet possible. Knowledge that a new world exists, rather than just a version of the current one, is manifest when what is possible – in this case the miracle – has already taken place. The manda therefore assumes critical importance, since it ensures that the change is both recognized and publically acknowledged.

The manda is critical and is treated as such by devotees. Inasmuch as it might be a modest offering or the performance of an extraordinary event of endurance, it must be practically doable. Of what a manda should consist is not defined. Personal limitations and identifications of self with place and with a particular Christ may function to orient and define the nature of the promises made by devotees in exchange for putting the world together in a new way.

This point about mandas is not trivial, since it is the primary mechanism for ensuring that new worlds are not made willy nilly. The manda is the limiting factor. One does not offer mandas one cannot or does not want to keep. To do so is to risk not only bad fortune but also punishment from the Christ.

What is at stake is subtle. Conceptually, a manda is not a repayment for what the Christ has done, since that would imply 'the doing' and 'the repayment' as congruent structurally and as taking place within a single world. Such framing would understand a manda to be a debt and a loss. Instead, a manda is a resource that is given. Each person who asks something of El Nazareno offers what is in their capacity to give and communicates that to the Christ and to themselves, through prayer and other devotional acts, the manda being chief among them.

Prayer and Place

Do you pray to the Black Christ, I asked Salome: 'No', she shook her head, 'not always, but I pray every night, well precisely to the Christ of Portobelo, if not to him to God himself, to the virgin and the angels. I have always done so since I was a little girl.' Salome's response focused my general characterization of the Black Christ in terms of place.

While there is no hard rule that I have been able to tease out for why people associate with or pray to one Christ rather than another, territorial proximity and one's sense of belonging are two considerations. Camarena, a man in his fifties who had been born and raised in Panama's Interior, related how his daughter was ill and needed an operation on her arm that he could not afford. At the time, he lived in Colón on Panama's Atlantic coast. Though he describes himself as a follower of the Christ of Atalaya located in the Interior, he asked the Black Christ to heal his child. The Christ did so and Camarena went to Portobelo to complete the reciprocal circle by paying a manda.

When I asked why he chose to ask the Black Christ for help, he replied that the Black Christ works and because visiting that Christ was more convenient than travelling to the Interior, especially when he was needed at home. At that time, Portobelo was accessible from Colón most easily by a short boat trip as opposed to over land. One could come and go in an afternoon, whereas to get to Atalaya and back would have meant undertaking a journey of several days.

Salome, though some thirty years Camarena's junior, echoed his views about convenience. When she lived in Colón with her family, she would appeal to the Black Christ if those close to her needed healing. She explained this was because everyone in Colón knows of the Black Christ, 'it is what is familiar and it is nothing to travel to Portobelo [from Colón]. There is a bus that regularly shuttles people back and forth between the two cities.'

Monica, a woman in her thirties raised in the capital but with a granduncle who had been a priest in the Interior and with whom she had spent considerable time as a girl, expanded on these ideas:

> You see one's devotion has to do with where you belong, where you come from, with the sentiments and feelings you have for a place. Where you have suffered. Here in Portobelo, we have the Black Christ for a mix of people. If you recall the celebration we attended here, people come here to Portobelo to make mandas from Veraguas or Chiriquí [provinces far removed from Colón]. People from the Interior feel welcome here. It seems to me that also there is a sentiment within the family here. It could be the effect of the Black Christ on the people who live here in the community. But the Black Christ is for all Panas [a nickname Panamanians use to describe themselves] that feel union with him and that believe in his miracles. Those that make a promise, or who are in a difficult situation. I think the people identify more with the Black Christ. You meet the image in taxis, in busses. You see it all the time in taxis. Some will carry the image stamped on a piece of wood or they will have a three-dimensional figurine. Something to always carry along. I have asked taxi drivers why they carry it and they have told me that they are devotees, and that they have been to Portobelo to pay a manda. One said he went to see if he is really a saint. And he found out that he is. He spoke of a miracle. I don't remember the details well but he said that he didn't have the money to attend and then the money just fell into his lap and he went by bus. He says that when he has money again, he will go with his entire family.

The link between the Nazarene, devotion and region is in a sense a shortening of the distance that one might have to travel to make contact with Christ in general or with the Black Christ in particular. But, it also suggests not only a lessening of the potential hardships of being on the road as a pilgrim (including time away from ones family as in the case of Camarena), but a shortening of the contemplative time that seeking to make contact with the Christ would require. In other words, when Christ is needed, often it may not be in a person's interest to be far away from where Christ could be found, because while Christ is everywhere, he is often more accessible in particular places.

Conclusion: Relapse and Return to a Place in the World

I was warned against making my first trip to Portobelo to see the Black Christ during the two-day October celebration. I was told by almost everyone I met in Panama City, including policemen, that the Black Christ was the patron saint of pickpockets, thieves and criminals and that I would do well to avoid Portobelo, the Black Christ and the people who worship him. The first few paragraphs of a Reuters wire story titled 'Criminals go on knees for pardon!' that was filed from Portobelo, 22 October 2005, sums up the perceptions that informed the comments and advice I was given:

> Thousands of muggers, burglars and drug dealers, some crawling on their knees, paraded through a Panamanian town on Friday, seeking forgiveness from Christ for their crimes. Wrongdoers lined up at the San Felipe church to pay homage to the Black Christ statue in an annual event to mark what many in Panama see as the patron saint's day of criminals. Up to 50,000 pilgrims, not all of them lawbreakers, made the 60-mile (96-km) trek to the Atlantic coast city of Portobelo from Panama City on foot. . . .
>
> 'I came here to ask the Christ to pardon my sins. And to thank him for getting me out of jail on a few occasions,' said one tattooed youth who gave his name only as Elisier. 'I robbed a Chinese store with a handgun. But now I have changed, with the Christ's help. I am a new man.'
>
> Reluctant to enter into the spirit of forgiveness, police set up checkpoints on roads into the city to see if they could nab wanted criminals and illegal immigrants among the crowds, but no arrests were reported.

The Christ and the promise of solution for whatever ails bring penitents to Portobelo annually, but also daily when the need arises and opportunity allows. El Nazareno reaches through the space created by his reputation as a healer to situate people and their problems in a place, which is Portobelo, and from that place they are to return to some other place, usually at a distance, where the myth and reputation of the Black Christ of Portobelo holds sway.

At some point, those who recognize their capacity to remake the world in which they live with the help of Portobelo's Christ must see El Nazareno and be seen by him. This is the only way to act on the logic of the manda, which is the performed, symbolic mechanism through which new worlds are instantiated by a miracle. That miracle is the tug of relapse and return. It is the possibility of conditions being other than they are. This is precisely what those who need healing as well as those who break the law are after (and acknowledgement of this logic is why the police set their traps for them where and when they do).

The idea of Christ as rooted in a place extends back to the European Renaissance and is often opposed to notions of progress and of being modern. In practice, I think the idea that Christ is associated with a place helps give rise to the identity of Christ as a multiple that is locally rooted and so necessarily modern. Pilgrims recognize the Black Christ as belonging to a given territory as well as being a part of them. El Nazareno then is not only property but the property of a place. The figure of Christ becomes rooted to a locale, which is but one of many.

One must also acknowledge the possibility that Christ as an image was multiplied and circulated and established in various locations so that pilgrims and penitents would not have to travel so far to benefit from the Christ's presence. This multiple Christ could also represent a place in much the same way as would a flag, and thus serve as a focal point for the expression of sentiments of belonging. It is almost paradoxical then that the Black Christ connects people in spite of their physical proximity to one another, and yet simultaneously reinforces the understanding of spatial distance as a barrier to veneration, thus helping give rise to a functional need for Christ as a multiple and for the performance of a manda to involve a journey and a return to a new world.

Though the image of the Portobelo Christ has become associated with the expressions and places of popular culture, as Monica described above, having been painted onto the sides of busses in the capital, incorporated into murals, painted onto blue jeans or worn as an image on oversized necklaces favoured by 90s hip hop artists, these very much remain representations of something that is in Portobelo. In Portobelo in October one can buy prayer cards, candles, images, rosaries, necklaces or any manner of objects with the likeness of 'El Naza'. But the statue itself is singular. It only exists in one location. Portobelo is its mythical centre. It radiates through images while remaining very much in-situ; in the location Salome says she understands the Christ desired to be.

The fungibility of its material form produces the context of the consumption of its miracles, wherever they might be located. In other words it produces miracles wherever it might be and it might be in many places. The Esquipulas Christs are said to have cured 'leprosy, blindness, muteness, insanity, paralysis, rabies, yellow fever, malaria, tetanus and haemorrhages. In earlier times, salvation in shipwreck and battle was also sought' (Hunter and de Klein 1984: 159). Similarly, the Christ of Portobelo is said to have cured both physical and mental ailments but in my own research he tends to be associated with recovery from physical illness. Unlike the Esquipulas Christs that have migrated to Panama from Guatemala and Mexico, El Nazareno is singular and that singularity suggests a relationship with place and thus stasis marks his identity in a significant way.

Panamanians often carefully explained to me that 'We are culturally Catholic, not religiously Catholic.' Being culturally Catholic means taking on the outward trappings of Catholicism – stating a belief in God, crossing oneself, maintaining holidays and some of their restrictions – yet without acceptance of Church doctrine. It is through the culture of Catholicism in everyday practice that the world is remade. It emerges through multiple symbolic acts of devotion and circulates through multiple images but its point of origin is always the same.

A miracle then is the process by which a possible world is opened through acts of devotion. In the case of the Black Christ of Portobelo, El Nazareno makes real the truth of possible worlds through requiring that devotees engage sickness as an event that not only courts possibility but requires iterative action. Operationalizing the logic of the 'manda' that initiates the work of worldmaking is to affirm the act of 'return' that is made possible and caused by 'relapse' into circumstances, relations and conditions that mark our lives as chronic conditions given to intervention and change. Mostly, the Nazarene in Portobelo deals with illness, but like the clan of saints to which he belongs, he also responds to the minutiae of everyday life.

In such a context, uncertainty is an index of life, thus defining the absolute truth-value not of what is already known as Csordas describes, but of the possibility of what has already been deemed impossible in the given profane world. This is why the miraculous exists not in that world that is given, but in the new world that can be made through the possible and real worlds that miracles bring into being. The Black Christ of Portobelo suggests that miracles require initiation into the process that acknowledges the completion of an act that may be called for again, because such is life. 'Relapse' and 'return'; 'return' and 'relapse': an ongoing oscillation out of which true new worlds emerge that from the vantage point of the old indeed appear to be miraculous.

Rodney J. Reynolds is a teaching fellow in global health and anthropology at the University College London's Institute for Global Health (IGH). He lectures and leads postgraduate and undergraduate seminars in medical anthropology, behaviour change and global health. Rodney's research interests include medical anthropology as it intersects with material and visual culture, belonging and politics of representation and well-being. His current research investigates barriers to cultural competency training for European physicians, obesity and how technology-led collaborative frameworks between universities and industry might encourage solutions to pressing global problems. He is a co-author of *The Lancet* Special Issue on Culture and Health.

References

Csordas, T. 1997. *The Sacred Self: A Cultural Phenomenology of Charismatic Healing.* Berkely: University of California Press.

Fortes, M. 1987. 'Coping with Destiny.' In Fortes, *Religion, Morality, and the Person: Essays on Tallensi Religion.* Cambridge: Cambridge University Press, pp. 145–74.

Garica, A. 2010. *The Pastoral Clinic: Addiction and Dispossession along the Rio Grande.* Berkeley: University of California Press.

Goodman, N. 1978. *Ways of Worldmaking.* Indianapolis, IN: Hackett Publishing Company.

Montenegro, L., L. Perdomo and X. P. de Castillo. 1983. 'En Torno a la Religiosidad Popular y al Cristo de Portobelo.' *Revista La Antigua* No. 21, Panamá.

Chapter 3

Madness and Miracles

Hoping for Healing in Rural Ghana

Ursula M. Read

Man seeks not so much God as the miraculous.
– Fyodor Dostoyevsky, *The Brothers Karamazov*

Hoping for a Miracle

In Ghana the predominance of Pentecostal churches in the healing marketplace is manifest in the large number of 'prayer camps' established throughout the country. Along the major roads brightly painted signboards advertise prayer camps with names such as Miracle Healing Centre, Mount Zion and The Blood of Jesus. Known in Twi[1] as *mpaeɛbɔ fie*, prayer home, the camps began to proliferate in the early 1990s as sites for 'deliverance' from demon possession (Gifford 2001; Meyer 1998a; Tetteh 1999: 23) and centre around a charismatic 'prophet' or pastor who claims healing powers. As described by one Pentecostal leader, 'Prayer camps are places where people who want healing, deliverance or a miracle for a special need in their life come, camp and pray' (van Dijk 1997: 148). Individuals and their relatives may stay for several weeks or months at the camps undergoing prayer, fasting and deliverance from evil spirits. Prayer camps and Pentecostal churches have become so popular in the treatment of mental illness in Ghana that research in Kumasi, the second city, found that more people had consulted a pastor than a traditional healer before coming for psychiatric treatment (Appiah-Poku et al. 2004). Such pastors explicitly draw on biblical precedent for miraculous healing and the casting out of demons. However, with a focus on confession and the forces of evil, their practices

Notes for this chapter begin on page 62.

also echo those of the shrines, and tap into a longstanding discourse on the transformative potential of supernatural powers.

This chapter draws on ethnographic research conducted with people with mental illness and their families living in and around Kintampo, a market town in the centre of Ghana. This involved repeated visits to people with mental illness and their families in household compounds, prayer camps and a nearby shrine famed for its treatment of the mad. Due to its history of trade and migration, the town contains a multi-ethnic mix of Akan Bono, Mo and various northern groups, with about 60 per cent Christian and 30 per cent Muslim.[2] The healing options for madness are equally diverse: beside the Pentecostal churches, many *akɔmfoɔ* ('fetish priests' or 'traditional healers')[3] also treat mental illness predominantly through the use of potent herbal mixtures, as well as rituals of confession and propitiation. In addition, Muslim *mallams* offer prayers and readings from the Qur'an to combat possession by *jinn*. Biomedical treatment with psychotropic medication, commonly called 'hospital medicine', is also available via a few community psychiatric nurses and a regional psychiatric unit. However, these local services are still relatively unknown and underused. More notorious are the 'asylums', the three large psychiatric hospitals a day's journey away on the coast.

The expectation of a miracle and the possibility of its dramatic demonstration stimulated hopes of an ultimate cure for the many ailments that troubled the residents of Kintampo. Tales of miraculous events witnessed or overheard added spice to everyday conversations and church sermons focused on the miracles of the Bible and the possibility of their reenactment before the eyes of the congregation. Many Pentecostal services ended with enthusiastic 'testimonies' from members of the congregation who recounted tales of their personal experience of the healing power of Jesus. However, the discourse of 'signs and wonders' was not confined to the Christian churches. Fantastic tales were also told of the exploits of powerful *akɔmfoɔ* who were able to turn paper into money and cut a man open, take out his intestines, eat them and then restore him to life, to quote just two accounts claimed to have been personally witnessed by the teller. Ghanaian popular media reported such stories in newspapers, TV and radio, and they were a source of fascination, entertainment and speculation, stimulating both admiration and censure regarding the moral foundation of such miraculous power.

William Arens and Ivan Karp (1989: xvi) argue that in African societies 'power does not emanate from a single source and social formations are composed of centers and epicenters of power in dynamic relationship with one another'. The positioning of pastors in relation to both *akɔmfoɔ* and hospital medicine, and of those who attend prayer camps vis-à-vis the competing claims of healers, is exemplary of this 'dynamic relationship'. In Ghana the demonstration of power (Twi: *tumi*) through the performance of miracles and the healing of sickness is crucial as evidence of supernaturally inspired authority (McLeod 1981: 67). *Tumi* is held to derive both from the *abosom* of the shrines and from the Christian God, who is sometimes addressed as *Onyamɛ tumfo*, powerful God. Emmanuel Akyeampong and Pashington Obeng (1995) translate *tumi* as 'the ability to bring about change'. Thus in healing *tumi* manifests in a change from sickness to health. In a universe suffused with *tumi*, those with knowledge, such as *akɔmfoɔ* and other ritual specialists, can tap into this power: 'access to power . . . was available to anyone who knew how to make use of Onyame's powerful universe for good or evil' (ibid.: 483). Today the manifestation of miraculous healing power is central to the popularity of Pentecostal churches and prayer camps. Since demons and witchcraft are often viewed within Pentecostal doctrine as causing sickness, healing takes place through the use of prolonged and intense prayer and forceful deliverance – demonstrations of the charisma of the healer and his power over the forces of evil.

However, as Akyeampong and Obeng make clear, *tumi* is fundamentally ambiguous and can be used to harm as well as to heal. Hence the source and legitimacy of healing power is subject to judgement, which in turn carries implications for its efficacy. Power may be rendered morally suspect if it is thought to have been obtained illicitly through the use of 'medicines' ('juju' or sorcery) and witchcraft. Thus, as Jack Goody describes in northern Ghana, 'Shrines are evaluated for what they can and cannot do, in terms of their strength and weakness, the degree of power or efficacy' (1975: 92). A similar process of evaluation was applied to the first Christian missions, as Africans deciding whether to convert looked for evidence of a superior power against sickness from the missionaries and their medicine (Goody 1975; Hawkins 1997; Peel 2000). As anthropologists since Evans-Pritchard have observed, a spirit of scepticism or doubt is inherent in this empirical approach – a god must prove its power or it will be discarded (Barber 1981; Goody 1975; McLeod 1981). This scepticism extended to the healing claims of Pentecostal pastors. As Paul Brodwin (1996) observes in Haiti, the search for healing in contexts of medical pluralism is suffused with judgements regarding the moral foundations of those who claim healing powers. One informant referred to 'Africans' and their 'deceptive tricks' when describing a fruitless and expensive search for a cure among healers who promised much but delivered little. Scurrilous talk abounded among Kintampo residents and in the media of '419' pastors (a reference to Nigeria's famous anti-fraud legislation) who made extravagant claims for their healing powers but were concerned only with making money and sexually exploiting gullible female congregants.

In this moral landscape, pastors positioned themselves as battling for good against the evil powers of *akɔmfoɔ*, in which the powers of the *abosom*, the 'small gods' of Akan religion, are aligned with Christian 'demons' (Meyer 1994). In this discourse the power of Jesus is greater than the powers of the Devil, and of the *abosom* and those who serve them. However this hierarchy of power is inverted by suspicions that some Christian pastors may obtain healing powers from *akɔmfoɔ*.[4] On a visit to the shrine, my companions pointed out with some titillation that the men consulting the *ɔkɔmfo* that afternoon were pastors, confirmed by the word 'clergy' emblazoned across the top of their car windscreen. On occasions, pastors are openly challenged to duels of strength by *akɔmfoɔ*. In 2008 Ghana's most notorious *ɔkɔmfo*, Kwaku Bonsam, publically invited the pastor of the Ebenezer Miracle church to such a duel, a challenge that made national headlines.[5] Nonetheless, as Evans-Pritchard (1976 [1937]: 107) observed, scepticism towards particular healers does not undermine conviction of the existence of spiritual powers and the ability of some to manipulate them for healing. Indeed, such doubt appears paradoxically to encourage the proliferation of novel cures, as witnessed in the exponential increase in Pentecostal and other healers in Ghana and throughout sub-Saharan Africa (Hampshire and Owusu 2013). Doubt may be directed against the diagnosis rather than the treatment, hence treatments remain valid despite a lack of success in individual cases (Last 1992: 401). Faith in the possibility of miraculous healing thus persists despite scepticism regarding individual pastors' motivations and the source of their powers.

In this chapter I consider how this discourse of spiritual power and miracles was taken up by people with mental illness and their families in the search for a cure, as well as by pastors who offered healing. Health workers would often lament that the psychiatric hospital was the 'last point of call', employed only after other avenues of treatment such as churches and shrines had been tried and failed. Campaigns urge families of those with mental illness to 'get treatment for them at the psychiatric hospitals and clinics before anywhere else', attributing 'ignorance' as to the nature of mental illness as the cause of families failing to seek medical treatment.[6] However, nearly all the informants I met with mental illness *had* used the psychiatric hospitals, sometimes as the first place of treatment, and often at more than one facility. It was only when hospital medicine had been tried and failed to achieve lasting change that people then turned or returned to shrines and churches. The structural failings of mental health services in many low-income countries, including Ghana, inform campaigns to 'scale up' the provision of mental health services, particularly to rural areas (Lancet Global Mental Health Group 2007). Certainly the lack of follow-up and the difficulty in obtaining ongoing supplies of medication may have been a factor in people discontinuing psychiatric treatment, as well as the poor conditions in the country's psychiatric hospitals. However, the limitations of psychiatric treatment itself also seemed influential in determining the course of action chosen by families and those with mental illness. Uncertainty regarding the cause of mental illness and its resistance to cure made mental illness particularly suitable to be taken up in the discourse surrounding the possibility of other sources of healing power. Rather than a consequence of 'ignorance', the continued search for healing in the face of the perceived failures of psychiatric treatment appears, I suggest, to be inspired by hope for an ultimate cure. Whilst this draws on a faith in the miraculous with roots in African religion, it is further nurtured and sustained by the promises of Christian pastors who draw on biblical images of Jesus as healer to position themselves as having access to exceptional healing powers and take the lead in a pluralistic and competitive healing marketplace.

Alice's Story

In July 2007 Kintampo came alive with talk of the 'Healing Jesus' crusade that, after months of preparation in local churches, was about to come to town. Under the auspices of Dag Heward-Mills, a medical doctor and 'bishop' of the Accra-based Lighthouse Chapel International, a Pentecostal 'mega church', the Healing Jesus crusade tours rural towns promising healing miracles for the sick. Two large articulated trucks parked up on a patch of empty ground emblazoned with photos of Heward-Mills alongside scenes of a woman pushing a wheelchair, bandages being removed and a biblical quotation: 'The blind see, the lame walk, the lepers are cleansed, the deaf hear, the dead are raised, to the poor the gospel is preached!' Excitement mounted as the opening night of the crusade approached and it seemed everyone in town planned to attend. In conversations regarding the impending event, speculation centered on this topic: Would we see a miracle?

On the morning after the opening night of the crusade, in which indeed many claimed to have seen the lame walk, I bumped into an elderly woman I knew. She had long complained of pain and swelling in her knee, and had come seeking a cure. However after we exchanged greetings, she told me she had also brought her daughter, Alice, for healing. She raised her arm and pointed outside. Alice's condition was unmistakable. She was walking wildly around the grassy area outside the tent, shouting and gesticulating, clearly *abɔdamfɔ*, a mad person. Alice had been ill for over ten years, and from the open compound of her mother's home in a densely populated quarter of Kintampo, her dramatic and shameful symptoms were witnessed by all around. When she had one of her periodic breakdowns, she refused to eat and did not sleep, singing, crying and shouting insults, often throughout the night.

The impact on Alice's family finances had been catastrophic. The costs of treatment had depleted all the savings that her mother had accrued over twenty-seven years as a kitchen assistant in a government post. At first her mother sought treatment at a 'miracle healing' church run by a renowned pastor, spending around the equivalent of US$150 on treatment with 'anointing oils'. Eventually they were told to leave the camp because Alice was not improving and her behaviour was disruptive. Alice's brother was critical of the pastor and believed he was motivated solely by material gain, a suspicion augmented by the number of cars and gold jewelry the pastor possessed. The family then sought treatment at a shrine where Alice was diagnosed as possessed by *mmoatia*, small forest-dwelling spirits. However, she was beaten so severely to rid her of these spirits that her mother took her away from the shrine fearing she would be beaten to death. Eventually a pastor from one of the churches they approached recommended that Alice's mother send her to one of the national psychiatric hospitals.

Alice responded well to treatment with antipsychotic drugs at the hospital and after a few months returned home, with instructions to continue her medication. The family, much relieved, held a thanksgiving service in the Catholic church. However, soon Alice stopped taking the tablets and became unwell again: aggressive, abusive and very disruptive to the household. This pattern repeated itself over the following years, resulting in four admissions to hospital over five years. Each time she was given medication and recovered. However, once she got home she refused to take it, complaining that it made her feel 'lazy'. It was unsurprising therefore that Alice's mother maintained her hope for a miraculous cure that would heal Alice's illness once and for all. Sadly there was no miracle for Alice that day at the Healing Jesus Crusade, but with a donation from the team for travelling expenses, and the help of her brother and some young men, Alice was sent back to the psychiatric hospital where she remained for many months. Her mother was fearful that if Alice returned home she would once again stop her medication and relapse. As she explained: 'What I was told at [the hospital] was that she has to take the medicine all the time. And if she takes it all the time it will help her recover from the illness. But if she doesn't take it and leaves some, then the illness can't go.'

Cooling but Not Curing

Talking *basabasa* (in a confused, disorderly and meaningless way), walking aimlessly, sleeplessness and aggressive and destructive behaviour were the paradigmatic symptoms of madness in Ghana. Hence madness, as with states of heightened emotion such as anxiety, anger and grief, is classified as 'hot'.[7] Heat is associated with 'nature' – dirt, disorder and the bush; coolness with 'culture' – sanity, cleanliness, order and social harmony (Mullings 1984; McCaskie 1995). A common symptom of mental illness or stress is that *tiri hyehye*, 'the head is burning'. Thus the methods employed in shrines and prayer camps, including herbal medicines, the use of mechanical restraints such as shackles and chains, beatings and prolonged periods of fasting, are seen to 'cool' madness and restore the person to the social order. However, as has been observed in anthropological studies of the 'indigenization' of pharmaceuticals in Africa (Bierlich 1999; Etkin et al. 1990; Kirby 1997; Whyte 1992), 'hospital medicine' (antipsychotics) was incorporated into this scheme, valued for its 'cooling' effect through its ability to induce sleep and weaken what some saw as the superhuman strength of the mentally ill. The sedative strength of antipsychotics, as with herbal medicine at the shrine, was respected as evidence of their power.

Despite these undisputed benefits, Alice was not alone in stopping her medication. The power of antipsychotics was ultimately felt to be *too* strong, compromising their efficacy, particularly in the long term. Like Alice, many mentally ill informants complained of the effects of 'hospital medicine', in particular feeling 'fat', 'lazy', 'weak' and tired. Gifty's mother explained that she had first taken her daughter to the hospital, however: 'When they gave her the medicine she became fine. But she became big [fat]. She was sleeping so much so she became very fat and I stopped and we took her to a prayer camp.' Importantly, many said that hospital medicine left them feeling too weak to work. In returning to everyday life, the sedative effects of antipsychotics seemed counter-intuitive where bodily strength is synonymous with health. The Twi *wɔ ahoɔden*, 'to have strength', carries the meaning of being healthy. Similarly *nya ahoɔden*, 'to get strength', means to be healed or to recover. Feeling strong was necessary not only for farming, but for many other routine daily tasks such as head-loading water, chopping wood, pounding fufu and carrying babies, all of which require physical exertion and endurance. Fatima had been given antipsychotic medication by the community psychiatric nurse in Kintampo, but refused to take it. Her father explained that since Africans have to work so hard they want medication that makes them feel strong and able to work, rather than the opposite. Indeed, medicines are valued precisely for their strengthening effect. Tonics and herbal medicines are often advertised for the strength or 'power' they claim to impart.[8]

The fact that on many occasions the illness returned when antipsychotic drugs were discontinued posed a further challenge to perceptions of their efficacy. The sister of Akosua, who had a long-standing mental illness, explained: 'I took her also to the hospital. At that time, I brought her here and they gave her medicine that will let her sleep, yet when the medicine gets finished, it also returns. Sometimes when they give her medicine it goes for about two months then it returns.' As Akosua herself told me: 'I want the one that will heal me completely so that it won't come back again.' Yet, as in Alice's case, hospital staff advised that antipsychotics had to be taken for months, even years, after the 'wild' symptoms of mental illness had died down in order to prevent relapse. Whilst other common illnesses such as diabetes ('sugar sickness') were known to require long-term and preventive hospital medicine, informants seemed to retain a hope of finding a medicine powerful enough to heal once and for all. Studies of help-seeking for chronic conditions such as diabetes and HIV/AIDS in Ghana similarly show that when confronted with the failure of biomedicine to eradicate the disease, many continue to seek a cure from traditional and Christian healers (Aikins 2005; Awusabo-Asare and Anarfi 1997; Mill 2001).

The persistent and recurring nature of madness troubled the families of those afflicted and raised suspicions of *sunsum yadeɛ*, 'spiritual disease', a form of sickness confirmed by its resistance to hospital medicine. Many of the classic studies of healing in Ghana have alluded to the distinction between 'illness of the body', *honam yadeɛ*, and 'illness of the spirit', *sunsum yadeɛ*.[9] Under such a schema madness is categorized as *sunsum yadeɛ* and linked to supernatural causation such as witchcraft, sorcery and curses (Field 1960; Fink 1989; Mullings 1984; Owoahene-Acheampong 1998; Warren 1982). However, *sunsum yadeɛ* does not equate with a disease for which no biological cause can be found, as in the psychiatric distinction between 'organic' and 'functional' mental illness[10]; it is suspected in any illness that is unusually prolonged, or 'keeps returning after treatment' (Konadu 2007: 164). The crucial marker of spiritual illness therefore lies not so much in its causation as in its persistence. Thus the distinction was more often made between 'spiritual' and 'hospital' sickness than between 'spiritual' and 'bodily' illness (Hill et al. 2003). In a study of healing for mental illness in southern Ghana Leith Mullings was told by a patient at a shrine: 'When the trouble has a spiritual cause, you may go to hospitals but they cannot help you' (1984: 96). Thus, as Brodwin (1996: 104–5) notes in the distinction between *maladi Satan* (an illness of Satan) and *maladi Bondye* (an illness of God) in rural Haiti, the failure of biomedicine is an important empirical marker in the aetiology of illness, encouraging an inquiry into other types of evidence. Indeed, throughout Africa and its diaspora the failure of an illness to respond after trying out a range of treatments has long prompted suspicions of 'something behind the illness'.[11] In this case both biomedicine and Pentecostal healing have been taken up into an approach in which the relative efficacy of any treatment carries implications for both establishing causation and the possibility of an ultimate cure.

The Power of Prayer

Pastors drew on this discourse to argue for the superiority of their methods in cases of intractable illness. During his sermon at the Healing Jesus Crusade, Heward-Mills explicitly set out the limits of hospital medicine – 'doctors cannot raise people from the dead' – and promised: 'If I give you Jesus Christ I give you something that is far greater than chloroquinine.'[12] A pastor illustrated how some illnesses cannot be cured by 'orthodox medicine' by citing the biblical story of the woman with 'an issue of blood' who 'had suffered many things of many physicians, and had spent all that she had and nothing bettered, but rather grew worse' (Mark 5:24–34). Like Heward-Mills, most pastors held to a view of illness as being due to *bonsam*, the Devil. Prophet Agyei, who was well known for treating the mad in his prayer camp, held that all sickness was ultimately spiritual: 'As for sickness, it is spiritual. In the scripture God did not create human beings to get sick. Most of it is due to Satan power, devil power, evil spirit which enters the person and makes them behave that way.' Furthermore, illness that is caused by 'devil power' is resistant to hospital medicine, as Prophet Agyei explained: 'When it happens like that no matter what amount of medication you give to the person, they cannot be cured.' Thus the relapsing course of severe mental illness and the limitations of antipsychotics acted to consolidate the diagnosis of demonic illness, and lent fuel to the claims of pastors to offer a more powerful treatment. This in turn reinforced a turn to the healing promises of pastors by the relatives of those with mental illness. Pastor Owusu, who also ran a prayer camp treating mental illness, positioned himself as offering a method of healing, 'hard' or strong prayers, superior to both hospital treatment and 'traditional medicine' in offering a potent and lasting cure:

> What I see about it is that most of the sicknesses are not doctor sickness. It is spiritual sickness. Those which are spiritual sicknesses, you'll go to the doctor but it won't work. You will go to a herbalist.[13] Some of the herbalists can work with spirits, but what I see about that aspect of sickness is that for spiritual sicknesses only very hard prayers can help you. Because some can heal your sickness but after a short time the illness comes back. Some of it comes like that, the spirit which is disturbing you, if you don't pray, the spirit won't leave. You will do and do, the person will never be healed. Unless you pray to remove that spirit, the person will never be well.

This narrative, of the superior power of prayer over the failures of hospital medicine, was repeated not only by pastors but also by those who gave 'testimony' to being healed. Such testimonies were an important part of healing services where people were invited by the pastor to come to the front of the church and recount the story of their miraculous recovery. A common structure of such testimonies was that doctors had been unable to stop the sickness, and the person had finally come to the pastor where they had found a cure. God had therefore done what the doctors could not. Such testimonies were greeted with rapturous applause, and, if the story was particularly dramatic, much excitement. Maame Grace, who ran another prayer camp that was growing increasingly popular with the families of those with mental illness, told the following story:

> Another person was at Holy Family Hospital, he was struggling. When I saw him, I knew it is not an illness that could be treated at hospital because some of the illnesses are not treatable at hospital. When he struggles then they would give him injection and he would sleep. So I said, if the doctor would discharge him, then they should bring him to me. When he came I prayed and saw that it is someone who wanted to kill him. So I drove all the family people away and it was left with the child and his mother and me and my children. And we prayed for him. Three days and he was well. He is attending school now.

This story exemplifies the common narrative structure of 'miracle healing'. First, the illness could not be treated at hospital; second, the identification of a malign spiritual cause, often, as in this case, witchcraft from within the family; third, the superior power of prayer; and finally the resumption of a productive and valued social role. The ending of this story – 'He is attending school now' – illustrated the goal of all healing that, it was often claimed, was not met by hospital medicine, i.e. the return of the person to a productive social role. The most distressing aspect of mental illness was the way in which it cut down a young and promising child in the family, often just as he or she was about to complete school, travel abroad or start a career that would in turn benefit all the family. In the words of informants, mental illness could render a child effectively 'useless' or 'wasted'. A return to work or school, and a reintegration into the domestic life of the family was thus the fulfilment of a family's hope for healing, a sign of health and wellness and not simply the removal of symptoms, which may be seen as the primary objective of psychotropic medication. Where health and adult status is allied with strength and social responsibility (Owoahene-Acheampong 1998), healing is signified by a return to productivity. The hope for complete and restorative healing contained in this narrative, with the stages of spiritual diagnosis, prayer and a return to full functioning, was repeated by the mother of Yaa, who had taken her to a prayer camp: 'When you go [to the prayer camp], they say someone wants to destroy the child so we'll pray that the person cannot destroy the child, so that she'll be healed and go back to school.'

Pastors thus aimed not for partial remission, but complete cure. As in the build-up to Heward-Mills's healing crusades, pastors explicitly nurtured the expectation of a miracle among their congregation by drawing on dramatic biblical examples as evidence for the superior power of Jesus, channeled through them as spiritually endowed 'men of God'. This allowed pastors to position themselves as having unique access to a power not available at the shrines of traditional healers, but, crucially, deployed to combat the same sources of evil. As James Pfeiffer (2005:258) observes of Christian healers in Mozambique who claim unique access to the healing power of the 'Holy Spirit': 'Since church healers ply the same spiritual terrain as local curandeiros, often exorcizing malevolent spirits using local terms and idioms, drawing this distinction becomes especially critical to attracting new members.' At one service at Maame Grace's prayer camp, her assistant pastor fervently proclaimed: 'You are going to see things here today you have never seen before' and recounted several biblical stories of miraculous healing. He then asked the congregation if they needed a miracle. Many raised their hands and came forward seeking healing. On his website Heward-Mills also advertised his claims to fulfil biblical precedent, even prophesy:

> Several authentic and supernatural occurrences have been documented in this doctor-turned-preacher's healing meetings which indeed confirm the scripture in Acts 2:22, '. . . a man approved of God among you by miracles, signs and wonders. . . .' At these crusades, the blind receive their sight, the lame walk, chronic diseases vanish, the demon afflicted have been delivered and even the dead have been raised.[14]

The biblical story of the Gadarene madman was exemplary for pastors dealing with mental illness. In his book *Demons and How to Deal With Them*, Heward-Mills draws on the authority provided by his medical training to draw explicit parallels between the Gadarene madman and the medical diagnosis of 'schizophrenia', which he defines as a 'demonic affliction'.[15] In recognizing that 'in spite of medical advances, these problems do not seem to go away' (2005: 17), Heward-Mills positions himself as offering a cure that is beyond the reach of medical science: 'There are still mad men like the mad man of Gadara and Jesus still heals such people' (ibid.). The story of the Gadarene madman was similarly cited by pastors in Kintampo as biblical precedent for the healing of the mentally ill. The biblical depiction of the madman who 'ware no clothes, neither abode in any house' (Luke 8:27), and who was 'often bound with fetters and chains' (Mark 5:4) resonated clearly with the cultural stereotype of aggressive and naked vagrant madmen, as well as being reflected in pastors' encounters with the most disturbed mentally ill who are often restrained with chains. Like Heward-Mills, pastors in Kintampo referred to the 'demon of madness', which could be cast out through 'deliverance'. Maame Grace repeated the tale of the Gadarene madman as evidence that mental illness could be caused by spiritual forces:

> The time that Jesus Christ was walking, he met a mad man. And the mad man asked him do you want to destroy us? And he asked him: 'Who are you?' and he said: 'We are legion.' That time it was spirit. And it was the spirit that made the mad man take something to cut himself. So at the moment he shouted at the spirit, it left and entered into the group of pigs. And he became normal. Therefore, it is spirit that inhabits the person. So when the spirit leaves, if he is naked, he sees that he has to wear clothes. He has to eat, he has to bath. If the spirit is in the person, he does not eat, he does not bath. If the spirit says: 'Stand there', he has to stand there for the whole day. He tells him to go here. Do this, do this. So it is spirit.

The practice of deliverance provides a dramatic enactment of the struggle between the power of Jesus and *Bonsam tumi* (Devil power), demonstrating, as one pastor put it, that 'the power of God is greater than the power of the Devil'. Deliverance was an intense physical and emotional performance with the pastors placing their hands forcefully on the person's head or other body parts whilst repeatedly shouting phrases such as 'Get out in the name of Jesus!' and describing violent images of 'breaking' and 'burning' the evil spirit. Sometimes the pastors clap their hands loudly over the person to drive the spirit out or hit the person's body with their hands. Those being delivered often appeared to resist, spinning around or staggering under the pastor's hands and then finally collapsing to the ground, 'slain' by the power of the Holy Spirit. Some spoke words in the voice of the possessing demon, refusing to leave, before admitting defeat before the superior power of the name of Jesus.

Prophet Agyei and Maame Grace took 'before and after' photographs of men and women who had attended their prayer camps, contrasting the 'before' shots, where they were shown in a stereotypical state of madness, in chains with matted or 'bushy' hair, their semi-naked bodies partially covered by torn and dirty clothes, with those taken after healing, showing the person neatly dressed in new clothes, their hair cut and styled, and released from chains. The dramatic contrast depicted in the photographs was shown as irrefutable evidence of a healing miracle, mirroring the Gadarene madman restored to being 'clothed, and in his right mind' (Luke 8:17–35). As Prophet Agyei explained, madness was the ideal condition for such photographic evidence since, unlike the common aches and pains for which people sought healing in Kintampo, madness was clearly visible in the ragged clothes and chains of those afflicted: 'These are the people when they are healed, it is wonderful. If someone has stomach ache and you cure the person, how can you tell? So these are the serious ones where everyone will see that – yes.' Likewise, a group of pastors from Kumasi visited Kintampo expressly to round up the vagrant mad around the town whom they publicly washed in disinfectant and dressed in clean clothes, displaying them afterwards to the enthralled spectators as miraculously healed. Once a mentally ill person had been healed in the prayer camps, the chains were removed, and the person was taken into the congregation to participate in the rituals of worship. The person often began to assist in the running of church services and to help farm the pastors' land or do other chores in the church compound. In such a way the person's restoration to health was demonstrated to all in the congregation, confirmed too through the ritual of verbal testimony.

When Healing Fails

Nonetheless, as in Alice's case, the search for a miracle often ended in disappointment and mental illness remained resilient to both hospital medicine and the healing claims of pastors. Kwasi's father's healing quest was typical. Having first tried the hospital and finding neither an explanation nor a lasting cure for his son's illness, he turned to a prayer camp:

> When the illness first came, and he was brought from where he was, I took him to Anakful [psychiatric hospital] for his brain to be examined. . . . They didn't explain anything, but prescribed some medicine for him. They told us that when the medicine is finished we should go to Sunyani [regional hospital]. When he took all the drugs we went back for more. Yet still, the illness was getting worse so we went to a prayer camp.

Kwasi's father explained that the family had chosen to go to a prayer camp in effect to rule out the possibility of evil spirits as a cause of Kwasi's illness: 'As for the prayer camps they say they can cast out evil spirits. No one knows if it is a spirit that is following him, an unclean spirit that is worrying him. That is why we took him there.' The pastor at the prayer camp had confirmed that it was indeed a spirit that was troubling Kwasi and that 'he will pray and as he continues praying the spirit will leave'. However, after a long spiritual battle over two and a half years the sickness returned: 'When we went there he [the pastor] also struggled with it. We even thought that the sickness perhaps because of the prayers all of it had perhaps been cast out and had gone. As it was the sickness came back again. . . . He couldn't drive the sickness away.'

Finally Kwasi's father turned to the shrine, where he expressed faith in the power of the *ɔkɔmfo*'s herbal treatment: 'As for the medicine, when you give it him the illness can pull out. . . . Because how it was like over there, the violence was too much. Now it has gone down a little.' He seemed to retain his optimism that a cure could be found and was pragmatic about the considerable time and money he had spent on the various healers he had approached. As he prepared himself to watch and wait on the effects of the *ɔkɔmfo*'s medicine, he observed: 'Now it is like we are throwing stones. If you hear: "Hey!" then you go there.' 'Throwing stones' is an idiom meaning to try things out. This 'try it and see' approach to help-seeking has also been observed in Uganda: 'Afflicted people "try out" (*ohugeraga*) a plan of action to see if it works', a process informed by 'a pragmatic attitude of experimentation' (Whyte 1997: 232). The outcome of the treatment, Kwasi's father explained, would then confirm the type of illness with which his son was afflicted. As he saw it, his son was afflicted with either *abonsam yadeɛ*, the devil's disease, or 'an illness which had come to kill him', or one which would 'come and then go'. Kwasi's father described the slow empirical process of 'watching and waiting' to see the effects of treatment:

> The illness came a long time ago so you need to watch the medicine used to treat him for some time. Maybe from six months to one year and if the treatment is not good . . . or the illness cannot be cured, then you can decide to look for a different one. Even if you go to the doctor and they give you medicine, you'll be given a lot of medicines to take. And if you finish taking them all and you don't get better, you will be given a different one when you go back. . . . So if you rush into seeking treatment in a day, it is not possible. As for disease, it comes very fast, but it goes little by little. Therefore if you continue taking the same medicine for about six months and still it does not improve, you will see. You will also see if any changes happen.

He contrasted the incomplete effects of hospital medicine to the complete cure promised by the churches and shrines: 'If a doctor gives you medicine which you take and your illness goes away but later comes back, then you are not healed. Maybe he just reduced it a bit.' Yet traditional medicine can also fail to fulfil its promise, as confirmed by other informants who had tried various healers without success. As Veronica Mhina (2009: 155) notes in Tanzania: 'Mentally ill people are let down by the promises of both bio-medicine and traditional healing systems which, though they may offer some temporary relief, seldom offer the chance of social acceptance or indeed a feeling of having really got better.'

Kwasi's father's approach exemplifies those who, like Alice's mother, retained hope for a cure despite the failures of particular doctors, pastors and traditional healers. In a pluralistic healing landscape, so long as one has the necessary resources, there is always the possibility of finding someone who possesses a greater power. Indeed, some argued that one might be healed despite the false promises of charlatan pastors through the power of faith alone. When I showed my young Catholic housemate a video recording of Maame Grace's healing service in which she dramatically claimed to cast out demons, she expressed disapproval, claiming that such persons get their powers from 'juju men'.[16] A pastor who ran a Pentecostal church that attracted many educated elite in Kintampo made a similar claim, despite practicing deliverance and healing himself, telling me that some pastors get their healing power from 'juju men' in exchange for human blood or body parts. Whilst this may produce short-term results, he explained, it does not last since the pact with the *ɔkɔmfo* must be constantly renewed through offerings, an arduous and expensive process. However, others, like my housemate, discounted the powers of pastors as the decisive factor in achieving a miraculous cure. If a miracle takes place, she argued, it is through the faith of the person who has come for healing, rather than through the pastor's power. She went on to describe a Nigerian movie she had seen in which a paralysed woman approached two 'fake' pastors for healing. To the pastors' astonishment she was healed. The message was that the miracle had been due purely to her 'faith'. This storyline reveals the underlying promise of 'miracle healing', which lies beyond human intermediaries in individual faith. Such faith frees spiritual power from human manipulation and moral ambiguity, allowing direct access to the divine source of healing.

Conclusion: 'God Will Do It'

Achille Mbembe (2002: 270) argues that in Africa the 'new religious imaginaire' is based on the mobilization of ideosymbolic formations, including 'the exercise of charisma (which authorizes the practice of oracular pronouncement and prophecy, of possession and healing)' and 'the domain of the miraculous (that is the belief that anything is possible)'. This latter was echoed in the songs sung at the Healing Jesus Crusade, 'I expect a miracle tonight, nothing is impossible to be impossible', and the biblical phrase 'All things are possible to him that believeth' (Mark 9:23). This 'religious imaginaire' appears to animate both sides in the search for healing, those who hope for a cure and those who promise to provide it. As Mbembe (ibid.) remarks, charisma, such as that exercised by healing pastors, invests the owners with a 'distinct, autonomous power and authority', one which provides the pastor with status as a 'man of God' who works to heal by means denied to the man of medicine. Thus 'the exercise of this authority places the thaumaturge in a hierarchical relationship with those who are not endowed with the same magic, the same know-how' (ibid.). The struggle for ascendancy in this hierarchy is thus fuelled by the discourse of spiritual power. Heward-Mills, as a qualified medical doctor, is able to exploit his expertise in both forms of 'magic' and 'know-how', thus captivating the hopes of thousands who attend his healing services. However, Mbembe (ibid.: 271) goes on to argue that in a culture of shortage and scarcity where access to 'desired goods' is restricted to a privileged few, people may attempt to access such goods through 'shadow interventions in the phantasmatic realm' and 'the powers of imagination are stimulated, intensified by the very unavailability of the objects of desire'. This references the Comaroffs' (1999) notion of 'occult economies' in which people turn to witchcraft and other occult means to obtain the wealth denied them by the structures of global capitalism. The same could be argued of the desire for health, in a context of medical scarcity, and the conundrum of diseases that remain resistant to biomedical intervention.

The poor quality and scarcity of psychiatric services in Ghana, as of all health services, particularly in rural areas, is testimony to the lack of investment in mental health care and its low priority despite years of promises and short-lived initiatives. Even those who do enter the psychiatric hospitals are unlikely to receive the highest standard of care due to a lack of qualified staff, overcrowded and unsanitary conditions and insufficient resources (Ofori-Atta, Read and Lund 2010). However, biomedicine faces challenges to its efficacy even in the best-resourced conditions. This is starkly evident in the treatment of 'severe and enduring' mental illness, including the psychiatric diagnoses of schizophrenia or bipolar affective disorder, which, as the term implies, can remain stubbornly persistent despite the best efforts of psychiatrists. Where, as in Ghana, the efficacy of a therapy is measured through pragmatic experimentation, the failure of psychotropic drugs to achieve a complete and lasting cure may further confirm the nature of mental illness as a spiritual disorder, beyond the reach of medical expertise. In this lacuna between the limitations of hospital medicine and the promises of spiritual healers, the imagination of those who seek treatment may be stimulated by an elusive 'object of desire', a miracle cure, able to address the ultimate cause of madness and restore the person to his or her social role.

Thus in approaching those who promise spiritual cures, families are not simply displaying 'ignorance' of mental illness as 'a medical condition requiring medical attention', as described in a report into human rights violations in the prayer camps (CHRI 2008).[17] Rather, the search for healing appears fuelled by hope for an ultimate cure, a hope that seems rarely extinguished despite years of disappointment and expense, nurtured by a faith in miracles and the boundless potential of supernatural powers. A long and unsuccessful hunt for healing did not deny the possibility of a miracle, as Alice's story shows. Rather, the failures of particular healers or specific cases could reinforce rather than dismantle the faith of those who sought healing, pushing families onwards in search of those with truly authentic healing powers. Scepticism was directed not towards miracles, and the power of God or gods to perform them, but towards those who claimed to access healing power, its moral source and the motivations for its use. As people in Kintampo comfort each other, *Nyame bɛ yɛ*, 'God will do it'. Though hope for a miracle may grow weak, the ultimate destiny of every mentally ill person is in God's hands. In the words of one father: 'You fought and fought, your strength is finished. What will you do? So now it is only God who will heal him.'

Ursula Read worked as an occupational therapist in UK mental health services before gaining a PhD in anthropology at University College London based on an ethnography of people with psychosis in Ghana. She is currently a postdoctoral fellow at the Centre de Recherche Médecine, Sciences, Santé, Santé Mentale et Société (CERMES3), Paris, France, working on an anthropological study of the emergence of rights-based approaches to mental illness in Ghana. Her research interests are in the social experience of mental illness and the ways in which global mental health interventions are constructed and translated on the ground.

Notes

1. Akan Twi is the most widely spoken language in Ghana and was widely used as a lingua franca in Kintampo. All terms in italics are in Twi.
2. Data taken from www.ghanadistricts.com.
3. *Akɔmfoɔ* (sing. *ɔkɔmfo*) are ritual specialists who practice *akɔm*, possession by the *obɔsom* (god) of the shrine. The *abɔsom* (pl.) instruct the *ɔkɔmfo* in the preparation and use of herbal medicines. Although Kintampo town itself no longer has a shrine, most of the surrounding villages contain a shrine and resident *ɔkɔmfo* who is consulted for protection, retribution and healing.
4. This doubt about the possible origins of power also extends to political figures in West Africa (Ellis and ter Haar 1998; Meyer 1998b; Strandsbjerg 2000). As Strandsbjerg (2000: 398) notes in the context of Togo: 'religion and politics can be considered as different ways of thinking about power'.
5. See 'Kwaku Bonsam: "I give pastors Kofi Kofi,"' http://ews.myjoyonline.com/news/200805/16071.asp.
6. This was the caption on a poster produced by BasicNeeds, a British mental health charity operating in Ghana. It accompanied a drawing of a semi-naked woman in chains, with a pastor standing over her holding a Bible.
7. The association of madness with heat has been observed in other African societies (Beneduce 1996; Whyte 1998).
8. Medicines, pills and tonics of all kinds are popular resources in Kintampo, taken not just for sickness, but to obtain physical strength (particularly to enhance sexual performance), spiritual power or protection. As elsewhere in Africa, medicines (sing. *aduro*; pl. *nnuro*) cover a wide range of substances used to both harm and heal including herbal preparations (*abibduro*, African medicine), *'asopiti nnuro'*, hospital medicine, and sorcery materials ('bad medicine' or 'juju' in English).
9. The *sunsum*, often translated as spirit, is the source of personality, character, disposition and intelligence (Konadu 2007). The *sunsum* is vulnerable to spiritual attack, for example, being consumed by witches.
10. 'Organic mental disorders' include dementia and delirium for which a clear organic pathology can be identified (WHO 1992). 'Functional' mental disorders include those such as schizophrenia or depression where an organic cause is less certain.
11. For example, Horton (1967: 60) describes how 'native doctors' will begin with a specific herbal treatment, and then if the illness does not respond will try another. It is only after various treatments fail that 'the suspicion will arise that "there is something else in this sickness"'. More recently Whyte (1997: 25–28) notes in Uganda a move from a 'symptomatic idiom' to an 'explanatory idiom' when a condition does not resolve itself.
12. Malaria is endemic in Kintampo. In fact the current treatment provided by the district hospital is the WHO-recommended combination of artesunate and amodiaquine. Malaria in Ghana is resistant to chloroquinine, which however remains widely and cheaply available.
13. The pastor here used the word *dunsini*, translated as 'herbalists', literally meaning 'one who works with parts of a tree'. Strictly speaking *dunsini* work with herbal medicines only and do not offer rituals to address spiritual concerns such as witchcraft (Konadu 2007). However *dunsini* is sometimes used interchangeably with *ɔkɔmfoɔ* or *obosomfo* (a priest who serves the *obosom*) to refer to those who are subsumed under the English 'traditional healers'. This is how it is used here.
14. Taken from the website of the Healing Jesus Crusade: http://www.healingjesuscrusade.org/evangelist.html. Heward-Mills has considerable influence among Pentecostal churches in Ghana due to the frequency of his large healing crusades, which penetrate

into the most rural areas of Ghana, the widespread distribution of his books, and his practice of offering scholarships to his Bible School for pastors from rural areas. His church, the Lighthouse Chapel International, has many branches throughout Ghana and the diaspora.

15. The naming of demons with medical diagnoses is a common practice within Pentecostal deliverance. Heward-Mills's linking of schizophrenia with demon possession echoes doctrines propagated by prominent Charismatic teachers in the United States whose books are widely read among Ghanaian Pentecostal pastors (Gifford 2001). Prophet Agyei showed me an American biblical concordance that stated: 'There are demon spirits for every sickness . . . known among men.'
16. My research assistant, however, who is a devout Pentecostal Christian, was impressed by Maame Grace's compassion and hard work, and expressed no doubts regarding her methods or her spiritual authority. This underscores the plurality of perspectives regarding the claims of healers and the subjective nature of people's responses.
17. The 'ignorance' of rural dwellers and an assumed belief in spiritual causation is widely cited to explain a failure to seek prompt medical treatment, as in the following assertion: 'The belief that certain diseases are spiritually caused and therefore should not be taken to hospital is based upon ignorance and superstition. The rejection of the germ theory of disease and resort to superstitious belief is responsible for many deaths in the rural areas' (Ohene-Konadu 1997: 334). In fact, rural hospitals and clinics are overwhelmed by people seeking treatment. However, a lack of medical expertise and resources as well as unsanitary conditions means that scepticism towards the efficacy of biomedical health care in rural and even urban areas in Ghana can often be well founded (see Horton 2001).

References

Aikins, A. D.-G. 2005. 'Healer Shopping in Africa: New Evidence from Rural-Urban Qualitative Study of Ghanaian Diabetes Experiences.' *British Medical Journal* 331(7519).

Akyeampong, E., and P. Obeng. 1995. 'Spirituality, Gender and Power in Asante History.' *International Journal of African Historical Studies* 28(3): 481–508.

Appiah-Kubi, K. 1981. *Man Cures, God Heals: Religion and Medical Practice Among the Akans of Ghana*. Totowa, NJ: Allanheld, Osmun and Co.

Appiah-Poku, J., et al. 2004. 'Previous Help Sought by Patients Presenting to Mental Health Services in Kumasi, Ghana.' *Social Psychology and Psychiatric Epidemiology* 39: 208–11.

Arens, W., and I. Karp. 1989. 'Introduction.' In W. Arens and I. Karp (eds), *Creativity of Power: Cosmology and Action in African Societies*. Washington, DC, and London: Smithsonian Institution Press, pp. xi–xxix.

Awusabo-Asare, K., and J. K. Anarfi. 1997. 'Health-seeking Behaviour of Persons with HIV/AIDS in Ghana.' *Health Transition Review*, supplement to vol. 7: 243–356.

Barber, K. 1981. 'How Man Makes God in West African: Yoruba Attitudes Towards the "Orisa".' *Africa* 51(3): 724–45.

Beneduce, R. 1996. 'Mental Disorders and Traditional Healing Systems Amongst the Dogon (Mali, West Africa).' *Transcultural Psychiatric Research Review* 33: 189–220.

Bierlich, B. 1999. 'Sacrifice, Plants, and Western Pharmaceuticals: Money and Health Care in Northern Ghana.' *Medical Anthropology Quarterly* 13(3): 316–37.
Brodwin, P. 1996. *Medicine and Morality in Haiti: The Contest for Healing Power.* Cambridge: Cambridge University Press.
CHRI. 2008. *Human Rights Violations in Prayer Camps and Access to Mental Health in Ghana.* Accra: Commonwealth Human Rights Initiative Africa.
Comaroff, J., and J. L. Comaroff. 1999. 'Occult Economies and the Violence of Abstraction: Notes from the South African Postcolony.' *Amercian Ethnologist* 26(2): 297–303.
Ellis, S., and G. ter Haar. 1998. 'Religion and Politics in Sub-Saharan Africa.' *Journal of Modern African Studies* 36(2): 175–201.
Etkin, N. L., P. J. Ross and I. Muassamu. 1990. 'The Indigenization of Pharmaceuticals: Therapeutic Transitions in Rural Hausaland.' *Social Science and Medicine* 30(8): 919–28.
Evans-Pritchard, E. E. 1976 [1937]. *Witchcraft, Oracles and Magic among the Azande.* Oxford: Clarendon Press.
Field, M. J. 1960. *Search for Security: An Ethno-psychiatric Study of Rural Ghana.* London: Faber and Faber.
Fink, H. E. 1989. *Religion, Disease and Healing in Ghana: A Case Study of Traditional Dormaa Medicine.* Munich: Trickster Wissenschaft.
Gifford, P. 2001. 'The Complex Provenance of Some Elements of African Pentecostal Theology.' In A. Corten and R. Marshall-Fratani (eds), *Between Babel and Pentecost: Transnational Pentecostalism in Africa and Latin America.* London: Hurst and Company, pp. 62–79.
Goody, J. 1975. 'Religion, Social Change and the Sociology of Conversion.' In J. Goody (ed.), *Changing Social Structure in Ghana: Essays in the Comparative Sociology of a New State and an Old Tradition.* London: International African Institute, pp. 91–106.
Hampshire, K. R., and S. A. Owusu. 2013. 'Grandfathers, Google, and Dreams: Medical Pluralism, Globalization, and New Healing Encounters in Ghana.' *Medical Anthropology* 32(3): 247–65.
Hawkins, S. 1997. 'To Pray or not to Pray: Politics, Medicine, and Conversion among the LoDagaa of Northern Ghana, 1929–1939.' *Canadian Journal of African Studies* 31(1): 50–85.
Heward-Mills, D. 2005. *Demons and How to Deal with Them.* Accra: Parchment House.
Hill, Z., et al. 2003. 'Recognising Childhood Illnesses and their Traditional Explanations: Exploring Options for Care-Seeking Interventions in the Context of the IMCI Strategy in Rural Ghana.' *Tropical Medicine and International Health* 8(7): 668–76.
Horton, R. 1967. 'African Traditional Thought and Western Science.' *Africa* 37(1): 50–71.
———. 2001. 'Ghana: Defining the African Challenge.' *Lancet* 358(9299): 2141–49.
Kirby, J. P. 1997. 'White, Red and Black: Colour Classification and Illness Management in Northern Ghana.' *Social Science and Medicine* 44(2): 215–30.

Konadu, K. 2007. *Indigenous Medicine and Knowledge in African Society*. London: Routledge.
Lancet Global Mental Health Group. 2007. 'Scale Up Services for Mental Disorders: A call for action.' *The Lancet* 370(9594): 1241–52.
Last, M. 1992. 'The Importance of Knowing About Not Knowing: Observations from Hausaland.' In S. Feierman and J. M. Janzen (eds), *The Social Basis of Health and Healing in Africa*. Berkeley: University of California Press, pp. 393–406.
———. 1993. 'Non-Western Concepts of Disease.' In W. Bynum and R. Porter (eds), *Companion Encyclopedia of the History of Medicine*. London: Routledge, pp. 634–60.
Mbembe, A. 2002. 'African Modes of Self-Writing.' *Public Culture* 14(1): 239–73.
McCaskie, T. C. 1995. *State and Society in Pre-Colonial Asante*. Cambridge: Cambridge University Press.
McLeod, M. D. 1981, *The Asante*. London: British Museum Publications Ltd.
Meyer, B. 1994. 'Beyond Syncretism: Translation and Diabolization in the Appropriation of Protestantism in Africa.' In C. Stewart and R. Shaw (eds), *Syncretism/Anti-Syncretism: The Politics of Religious Synthesis*. London: Routledge, pp. 45–68.
———. 1998a '"Make a Complete Break with the Past": Memory and Postcolonial Modernity in Ghanaian Pentecostalist Discourse.' *Journal of Religion in Africa* 28(3): 316–49.
———. 1998b. 'The Power of Money: Politics, Occult Forces and Pentecostalism in Ghana.' *African Studies Review* 41(3): 15–37.
Mhina, M. A. 2009. 'Coping with Mental Distress in Contemporary Dar Es Salaam.' In L. Haram and C. B. Yamba (eds), *Dealing with Uncertainty in Contemporary African Lives*. Stockholm: Nordiska Afrikainstitutet, pp. 141–58.
Mill, J. E. 2001. 'I'm not a "Basabasa" Woman: An Exploratory Model of HIV Illness in Ghanaian Women.' *Clinical Nursing Research* 10(3): 254–74.
Mullings, L. 1984. *Therapy, Ideology and Social Change: Mental Healing in Urban Ghana*. Berkeley: University of California Press.
Ofori-Atta, A., U. M. Read and C. Lund. 2010. 'A Situation Analysis of Mental Health Services and Legislation in Ghana: Challenges for Transformation.' *African Journal of Psychiatry* 13(2): 99–108.
Ohene-Konadu, K. 1997. 'Regional and Rural Inequalities in the Provision of Health Care in Rural Ghana.' In E. Kalipeni and P. Thiuri (eds), *Issues and Perspectives on Health Care in Contemporary Sub-Saharan Africa*. Lampeter: Edwin Mellen Press, pp. 333–43.
Owoahene-Acheampong, S. 1998. *Inculturation and African Religion: Indigenous and Western Approaches to Medical Practice*. New York: Peter Lang.
Peel, J. D. Y. 2000. *Religious Encounter and the Making of the Yoruba*. Bloomington: Indiana University Press.
Pfeiffer, J. 2005. 'Commodity *Fetishismo*, the Holy Spirit and the Turn to Pentecostal and African Independent Churches in Central Mozambique.' *Culture, Medicine and Psychiatry* 29(2): 255–83.

Strandsbjerg, C. 2000. 'Kerekou, God and the Ancestors: Religion and the Conception of Political Power in Benin.' *African Affairs* 99: 395–414.

Tetteh, J. N. 1999. 'The Dynamics of Prayer Camps and the Management of Women's Problems: A Case Study of Three Camps in the Eastern region of Ghana.' Unpublished MPhil thesis, University of Ghana, Legon.

Van Dijk, R. 1997. 'From Camp to Encompassment: Discourses of Transsubjectivity in the Ghananian Pentecostal Diaspora.' *Journal of Religion in Africa* 27(2): 135–59.

Warren, D. M. 1982. 'The Techiman-Bono Ethnomedical System.' In P. S. Yoder (ed.), *African Health and Healing Systems: Proceedings of a Symposium*, pp. 85–105. Los Angeles: Crossroads Press.

WHO. 1992. *ICD-10 International Classification of Mental and BehaviouralDisorders: Clinical Descriptions and Diagnostic Guidelines*. Geneva: World Health Organization.

Whyte, S. R. 1992. 'Pharmaceuticals as Folk Medicine: Transformations in the Social Relations of Health Care in Uganda.' *Culture, Medicine and Psychiatry* 16: 163–86.

———. 1997. *Questioning Misfortune: The Pragmatics of Uncertainty in Eastern Uganda*. Cambridge: Cambridge University Press.

———. 1998. 'Slow Cookers and Madmen: Competence of Heart and Head in Rural Uganda.' In R. Jenkins (ed.), *Questions of Competence: Culture, Classification and Intellectual Disability*. Cambridge: Cambridge University Press, pp. 153–75.

Chapter 4

'Sakawa' Rumours

Occult Internet Fraud and Ghanaian Identity

Alice Armstrong

'Sakawa' hit Ghanaian news headlines in 2007, prompting a nationwide epidemic of rumours that continue today. These rumours accuse young men of manipulating evil occult powers to perform successful internet fraud. In order to gain occult powers 'Sakawa boys' are said to perform socially grotesque rituals ranging from sleeping in coffins to cannibalism. These rituals endow Sakawa boys with the power to spiritually enter the internet and possess the mind of foreign fraud victims to extract quick and easy money. This immoral behaviour has consequences: Sakawa is said to end in infertility, mental illness, sickness or death for the perpetrators and the victims of their ritual sacrifices.

This is a contemporary ethnography, but the phenomenon is not completely new. Supernatural manipulation of the internet has continuities with long-standing West African cultural archetypes surrounding occult power and wealth that is gained at the expense of others. This expense is not just paid by family or friends; misfortune is inflicted on the entire nation. A wide range of Ghanaians condemn Sakawa as behaviour that is 'not Ghanaian' and that raises fears for Ghana's international reputation. An ontogenic approach is used to explore this change in scale, discussing the intimate relationship between identity boundaries and occult beliefs, which shift and change as Ghana develops as a nation. Sakawa rumours draw on long-standing occult idioms that negotiate 'Us' from 'Them' and are revealed as the latest renegotiation of a specifically Ghanaian identity, under and aware of the gaze of the world.

Notes for this chapter begin on page 86.

What is Sakawa?

> Sakawa boys sleep in coffins or don't wash for weeks; some even kill a small girl and eat her like fu fu [a national dish]! They do whatever these Sakawa leaders tell them to do and then they have the ju-ju power to do their evil tricks. Then their spirit can enter the internet, possessing the obruni (white person) to get their money! It is evil, greedy behaviour and they bring shame to Ghana.
> – Kwame, Abiriw, Ghana, 7 August 2009

Since 2007 Sakawa rumours have been fervently discussed throughout Ghana, engaging a wide range of people in their common condemnation of the behaviour. Sakawa is the use of evil occult powers to commit successful internet fraud, possessing the mind of the foreign cyber target. Ghanaian's definitions of Sakawa are flexible, involving the terms 'witchcraft, 'blood money' (known as 'sika duro' among the Akan), 'ju-ju' or 'magic' – and most commonly a combination of them all. Akan cosmology involves a complex array of spiritual beliefs and activities, many of which allow for spirit agents to impact on health, illness and the body. These spiritual misfortunes can include curses, punishment from the Supreme Being, the Devil and witchcraft (Opare-Henaku 2013). Annabella Opare-Henaku states, 'Akan, the largest ethnic group in Ghana, is noted for the use of supernatural attributions for various health-related issues. The supernatural attributions are based on Akan ontological belief that the universe is unitary such that there is no clear distinction between physical and spiritual occurrences' (ibid.: vii). As Sakawa combines illicit fraud and illicit occult activity, the behaviour has multifaceted consequences.

Sakawa is predominately rumoured to be the practice of young men aged 16–30 who gain their spiritual powers by joining secret and sinister Sakawa cults. These cults are thought to be led by wayward spiritual men, such as pastors or fetish priests, who hold illicit Sakawa meetings in the middle of the night. Initiations into such cults are rumored to involve socially grotesque acts such as sleeping in coffins, public nudity, refraining from bathing, ritual murder and cannibalism.

Sakawa boys are immorally harnessing occult power, which incurs a 'spiritual debt' requiring frequent secret ritual repayment. To not repay or maintain their spiritual contract results in loss of wealth, sudden bereavement, ill health and/or death. Strange antisocial behaviour might be attributed to someone's involvement in Sakawa activity, as they are 'driven' mad by their repetitive repulsive rituals. Opare-Henaku's informants suggested Sakawa as a potential cause for mental illness as a form of 'spiritual indulgence' whereby deliberate illicit engagement with the spirits could backfire and impact the perpetrator's health (ibid.). This is not about interaction with the spirits per se; well-being and good health can also be related to the spirits (see Read, this volume). However, as Sakawa is immoral occult interaction to gain wealth at the expense of others, it therefore incurs consequences.

There are accusations of Sakawa boys embroiling others in their rituals, causing physical and mental suffering. Rumours involve gaining occult power by stealing a girlfriend's menstrual pads to sacrifice her fertility, turning into snakes and biting victims who become unwell or die or having sex with women who then become mentally ill. The purpose of these rituals is to endow the internet fraudster with power to possess the mind of the foreign target, forcing them to hand over their money. This power can come from magic rings, handkerchiefs or enchanted laptops that allow the Sakawa boy to spiritually enter the internet. 'Kof-I', a self-proclaimed Sakawa boy who I chatted with via a Sakawa Facebook group,[1] boasted of his 'magic handkerchief', which he placed on the computer during internet scams, allowing him to 'enter the head' of his cyber victims. Sakawa boys are also rumoured to gain money by shape-shifting into snakes, which then vomit copious amounts of banknotes. The vomited Sakawa money is consumed conspicuously on goods such as 4×4s and is rarely shared. Such immoral greed angers many Ghanaians, however it is Sakawa's repercussions on Ghana's international reputation and future that seem to be particularly worrying.

The predominant theme during fieldwork was Ghanaian's condemnation of Sakawa as a threat to their nation's health and prosperity. This fear is also articulated in the media, with newspapers often carrying dramatic headlines such as 'Sakawa is ruining Ghana'.[2] The Christian media was particularly vocal in its condemnation of Sakawa, highlighting Christianity's moral obligation to reform the youth, who are often portrayed as epidemically immoral.[3] Sakawa was a common theme during church services, with pastors preaching of its evils and threat to Ghana's future.

An Anthropological Approach

After such an introduction I am cautious about a sensationalism of alterity that obscures the everyday ubiquity of witchcraft in Africa (Olivier de Sardan 1992). Sakawa is one form of a wide complex of Ghanaian occult beliefs and, although Sakawa rumours are contemporary, they are not a completely new phenomenon. However, anthropologists should also be aware of over-domesticating witchcraft rumours and explaining them away with cultural relativism (Geschiere 1997: 216). The occult may be omnipresent, however a perceived 'epidemic' is not 'normal' but is often a cause of fear and sometimes violence. Many expressed fears of becoming a victim of a Sakawa ritual, losing a limb, their life, fertility or mental health. Unlike some witchcraft in Ghana (Adinkrah 2004), Sakawa rumours do not result in violent witch-hunts. However, witchcraft rumours are not just a tool for controlling social concern (Gluckman 1963; Marwick 1964); they must also be valued as a cause of concern in themselves.

It will be these perceptions and feared consequences of Sakawa that will be the focus of this discussion, rather than the practice itself. As in Jeanne Favret-Saada's (1980) ethnography on witchcraft in the Bocage region of France, the focus will be on the power of words to make the witch – or in this case the Sakawa boy. Rather than explaining occult beliefs 'away', attention is given to the processes by which they exist and are maintained in social space over time, appreciating fears and concerns (ibid.). These fears are both a national concern and a national phenomenon, especially as Sakawa rumours and mass media coverage engage a wide range of Ghanaians. Sakawa is a popular topic of conversation among men and women, young and old, professionals and the unemployed.

Benedict Anderson's (1991) treatment of capitalism and communication technologies and their relationship to 'imagined communities' and nationalism proves more fruitful than 'witchcraft and modernity' inferences of capitalism as a juggernaut (Comaroff and Comaroff 1993 and 1999; Geschiere 1997: 135; Moore and Saunders 2001). Although the internet is somewhat emblematic of 'modernity' and is certainly surrounded by Ghanaian ambivalence, this discussion will be wary of treating Sakawa solely as a reaction to contemporary capitalism. Such an approach can seem to overemphasize a 'resistance to modernity', potentially reducing complex witchcraft beliefs to a 'snapshot' clash between traditional and modern (Shaw 2002).

Rosalind Shaw's work on witchcraft and memories of the slave trade in Sierra Leone (2002) illustrated that the relationship between past and contemporary occult beliefs is fluid. Shaw avoids a dichotomy of traditional versus modern, and this is how Sakawa will be contextualized. Interesting parallels are raised between the potentialities of internet technology and Ghanaian occult beliefs and values of power. Through these continuities longstanding occult idioms for negotiating immoral from moral, 'Us' from 'Them', will be revealed as interacting with contemporary ideas of nationalism and the concerns that this raises. As the discussion develops more questions arise: If Sakawa is a source of shame for Ghanaians why it is so freely talked about and splashed all over the media? If Ghanaians are so concerned with their global image, why are Sakawa rumours enthusiastically shared with a visiting British anthropology student? Points of concern and ambiguity are popular and important topics of conversation but it is also their openness that is interesting. Sakawa discourse is revealed as integral to the renegotiation of a Ghanaian identity. This renegotiation is a constant process, as boundaries and concerns change and develop over time with relevance to the past, present and future.

National Phenomenon, National Concern

Since 2007 Sakawa rumours have featured strongly in the Ghanaian media as well as inspiring popular entertainment such as 'Hip Life' songs and the Ghanaian film industry. Three months of Sakawa fieldwork and internet research included 55 newspaper articles, 27 television reports and 165 radio references, with many informants citing the radio during interviews. The radio is a particularly prominent means of spreading of Sakawa rumours due to its easy accessibility compared to the cost and level of literacy associated with television and newspapers. Radio as a medium for rumour is also unique in its relationship to African values surrounding oral history and informal communication networks (Ellis 1989: 321). Stephen Ellis uses the term 'pavement radio', translated from the French *radio trottoir*, to describe such networks as important sources of rumour and news (ibid.). These networks are constituted by people listening to the radio, talking about the radio and talking to each other. This casual nature of 'pavement radio' is not dismissed as trivial. Instead Ellis values gossip and rumour as featuring items of historical importance and long-time social interest such as sorcery (ibid.: 330; DiFonzo and Bordia 2007: 21; Shibutani 1966). The soundtrack to Ghanaian Tro tro (public minibus) journeys is often the radio and conversations among passengers were a rich resource for Sakawa rumours, drawing in people from different areas and backgrounds. Radio reports of the latest Sakawa rumour could provoke a chorus of tut-tuting, with passengers often engaging in conversation and exchanging shocking rumours. Widespread media coverage is therefore a cultural externalization whereby the same Sakawa rumours are discussed nationwide, creating a durable storage of meanings from which increasingly complex and sensationalist rumours propagate (Hannerz 1992: 27).

Ghana's diverse population is therefore united in a national Sakawa discourse, creating what Benedict Anderson has termed an 'imagined community' and a national identity. Anderson defines national identity as an emotional sense of belonging and loyalty, influencing a sense of one's self and an awareness of one's place within the world in relation to others (1991: 10). His 1991 book *Imagined Communities* analyses the fundamental role of the printing press in the construction of national identity. Accessible books and pamphlets created a collective consciousness whereby citizens were aware and relating to each other on an increasingly large and anonymous scale (ibid.: 6, 40; Eriksen 1993: 106). Ultimately, a common discursive medium was created as people read the same information without direct contact with one another. Although Ghana already has a constructed national identity, it is constantly under negotiation, and Sakawa rumours form a platform for Ghanaians to discuss how they see themselves and their society, both in the past and in the future (Hannerz 1987: 547). As people contribute and interact with the national media, they are constructing themselves as Ghanaians: 'You see this? [Points to article in the newspaper] These Sakawa boys, what next! It is evil and now look, it is ruining Ghana. What will people think? This is not Ghanaian behaviour, it's not right!' (from a conversation with Mrs. Quansah whilst reading the *Daily Graphic* and pointing at E. Yamoah's 'Mob battle police over 'Sakawa' boys', 26 July2009). Whether it's said between friends, in a national newspaper article[4] or on an internet message board,[5] Mrs. Quansah's indignant comment that Sakawa is 'not Ghanaian behaviour' is a common reaction throughout the nation. A sense of national identity stresses cultural similarity and solidarity, simultaneously drawing boundaries and creating others, and rumour often has a fundamental role in social boundary maintenance (Eriksen 1993: 6; Gluckman 1963: 308; Stewart and Strathern 2004: 30). Sakawa rumours and their portrayal of young men performing socially grotesque tasks constructs a dichotomy between 'Us' and 'Them', 'Moral and 'Immoral' and, ultimately, 'Ghanaian' and 'not-Ghanaian' behaviour (Douglas 2002: 27).

Interestingly, Sakawa is not only 'not Ghanaian' but is widely described by Ghanaians as 'Nigerian behaviour'. Ghanaians commonly believe 'Sakawa' to be a Hausa word and although 'Sakawa' itself is not Hausa,[6] it is the perception of it as a Nigerian word that is most interesting. Sakawa also has particular connotations with Nigeria's infamous '419 boy' internet fraudsters, with '419' being a reference to the paragraph in the Nigerian Criminal Code for fraud.[7] Eriksen has identified stereotypes as fundamental to the construction of a contrasting and often superior national identity, and Sakawa is not only discussed as evil but as specifically influenced by a stereotypical Nigeria (1993: 4). Ghanaian stereotypes of Nigerians as immoral, corrupt and violent have a long history. For example, the 1980s saw mass expulsions of each other's migrants, and Ghanaians returned with tales of xenophobic Nigerians, commonly stereotyping them as pompous, immoral and corrupt (McCaskie 2008: 332). Despite contemporary co-operative politics such as the Economic Community of West

African States (ECOWAS), popular relations remain strained and stereotypes continue (Gocking 2005). Friction is evident in an online September 2009 Ghanaian news report citing Nigerian Minister Godwin Abbe's comment: 'the volume of water generated in Ghana is not enough to flush the toilets in Lagos state'. Such insinuations of insignificance prompted Ghanaian journalists to retaliate with descriptions of Nigerians as corrupt and incompetent, whose flushing toilets probably do not work anyway (Owusu 2009).

However, the primary concern discussed in Sakawa rumours is the global scale of Nigeria's negative stereotypes surrounding corruption, drug trafficking and organized crime networks (Jordan-Smith 2007). As Daniel Jordan-Smith states: 'The infamous Nigerian email scams are emblematic of Nigeria's worldwide reputation for corruption' (2007: 28). Transparency International annually publishes a Corruption Perception Index (CPI) for public sector corruption in 180 countries. In 2009 Ghana ranked at 69, whereas Nigeria ranked at 130 (Transparency International 2009). Ghanaians are aware that Sakawa could have a similarly negative impact on Ghana's global reputation. Phillip Mayer once described witches as 'a traitor within the gates' (1954: 66), and many Ghanaians discuss Sakawa boys as betraying the values of their nation (Crawford 1967: 321; Douglas 2002: 127; Stewart and Strathern 2004: 67).

Sakawa boy's betrayal of their nation is felt particularly strongly as many Ghanaians are proud of their values and worldwide image. As the first African nation to gain independence in 1957, Ghana has maintained some of its 'leading light' status (Armah 1974: 175). US President Barack Obama chose Ghana as his first visit to Africa, which occurred during my fieldwork and prompted widespread national pride and elation.[8] With democracy and peace, Ghana is often put on a pedestal as a model modern African state, and Ghanaians are aware of the contrast to other African states, including Nigeria: 'It gladdens my heart to note that more Ghanaians enjoy electricity, potable water, functional health care, security, education and have a better standard of living that Nigerians, according to the United Nations' (Owusu 2009).

However many Ghanaians also believe that they are living in a time of increased danger, perceiving their nation as becoming increasingly similar to a stereotypical Nigeria with fears of epidemic immorality among the youth. One evening during fieldwork a small girl went missing and locals panicked, fearing the girl's abduction for Sakawa ritual murder. After a few hours the girl was found and her father reflected on his fear, articulating concerns surrounding a changing Ghana: 'With this Sakawa you have to watch out, now we cannot just let our children wander anymore' (fieldwork, 27 August 2009). During fieldwork the media also reported increasing gun crime and banditry throughout the nation.[9] As with Sakawa, Tom McCaskie's work on gun crime in Kumasi found that such dangerous and immoral behaviour is equated with Ghana becoming increasingly like Nigeria (2008: 437). A comment on a Sakawa online message board read: 'Ghana is in the process of being incurably infested with Nigerian crimes',[10] a sentiment that was often expressed by informants during my own interviews.

Interestingly, Sakawa rumours began in 2007, which was the same year that oil was found off Ghana's southwestern coast. It may be that Sakawa rumours emphasize Nigeria's perceived immoral influence because Ghanaians are increasingly negotiating concerns similar to Nigeria around the ambiguities of oil wealth (McCaskie 2008b). Ultimately, Ghanaian's stereotypes surrounding Nigeria are integral to a sense of national identity, a boundary that Sakawa boys worryingly transgress. This raises concerns for Ghanaians on both a national and international level, as they negotiate their past and their future with ideas of progress that particularly concern the youth.

The Youth and the Future: Social Reproduction Gone Wrong

Sakawa rumours are concerned with the youth and the power they embody. Sakawa is a practice of the young and Ghana has a large youth population, with the nation's median age at only 20.7 (CIA World Factbook 2010). Nationalism, like kinship, involves a sense of protecting vitality and the transmission of values across generations (Grosby 2005: 120). As A. L. Epstein stated in his analysis of rumour, to be talked about is a measure of social importance (1969, cited in Bleek 1976: 527). As the youth reproduce biologically and socially, they are the future and are therefore socially important. Sakawa rumours depict young men destroying this potential by acting immorally, but this is also meant literally, with rumours of Sakawa boys sacrificing their own and other's fertility in exchange for occult powers.

Many rumours also depict young men shape-shifting into snakes which then vomit money, an image which has also been linked to traditional Ewe symbols of often diabolical fertility (Meyer1995: 818; Wendl 2007: 16). Fertility, reproduction and witchcraft accusations are frequently related (Apter 1993; Auslander 1993; Jordan-Smith 2001a; Meyer 1995: 24). Wolf Bleek has written of witchcraft accusations against older women in Kwaku, Ghana, who are believed to inflict infertility on the younger generation. Citing the work of Siegfried Nadel (1952), Bleek interprets accusations as a resentment of women's power as a life-giving force (1976: 538–39). Sakawa is unusual, as accusations descend generational gradients; it is the youth who are sacrificing their own fertility, and therefore their own life-giving force, in return for occult power.

Ghanaians are particularly disturbed by this sacrifice, as Sakawa boys are rumoured to be educated young men who have a relatively high level of English and computer literacy. Sakawa boys are therefore portrayed as particularly greedy, not because they come from impoverished backgrounds but because they desire the glamorous lifestyle and camaraderie of the cult.[11] This selfish behaviour has repercussions for Ghanaian society. Their education means that Sakawa boys embody potential for themselves, their families, their communities as well as their nation – a potential that is destroyed by their illegal occult activities. Sakawa is thus feared for its reversal of expected behaviour and interruption of social reproduction, with consequences for the entire nation's health and prosperity.[12]

However, education in Ghana does not guarantee employment and there is an ever-increasing group of unemployed, educated young men looking for 'white-collar work'; in 2008 unemployment was 20 per cent (CIA World Factbook 2010). Misty Bastian has written of similar circumstances in Nigeria, with anxieties surrounding jobless, educated young men who are rumoured to form occult cults (Bastian 2001). These Nigerian cults are believed to be involved in criminal networks such as drug trafficking. As in Sakawa rumours, young men are said to perform grotesque rituals such as human sacrifice, which consequently endows members with occult powers such as mind possession (ibid.: 80). Nigerians fear the consequences that such behaviour might have on their society as a whole, as a powerful generation of educated but immoral and violent youths is created (ibid.: 81). As with Sakawa, there is a fear of social reproduction gone wrong.

'To Be a Man Is Not Easy': Sakawa, the State and Global Inequalities

In Ghana the government is frequently held responsible for Sakawa's context and the perceived epidemic of occult immorality. Politicians are criticized for failing to resolve economic problems and unemployment, thus making Sakawa a popular alternative. Promises of development are often unmet and Ghanaians are critically aware of national funds being siphoned off to politicians and elites: 'This Atta Mills [president at the time of fieldwork], he lies and lies. He promised to stop the suffering and now look . . . young men turn to Sakawa and crime. You must provide jobs for people or what will they do?' (Sebastian, Koforidua, Ghana, 2 August 2009).

Throughout West Africa it is usually politicians who are portrayed as greedy and cannibalistic, whereby 'eating' refers to food but can also be a metaphor for money and sex – i.e. life sources and resources (Bayart 1993; Geschiere 1997; Rowlands and Warnier 1988). Jean-François Bayart's term 'politics of the belly' therefore takes the phrase in both literal and figurative senses to discuss African attitudes to state nepotism (1993). However, with Sakawa it is young men who are rumoured to be cannibalistic, self-cannibalizing their own potential and therefore that of their nation. In the context of high expectations and restricting situations it is perhaps not surprising that young men are the focus of Sakawa rumours. Mary Douglas stated that witchcraft accusations often occur where 'roles are undefined, or so defined that they are impossible to perform' (2003: 127). This tension is evident in the popular Ghanaian male lament and graffiti slogan 'to be a man is not easy'.

As Tom McCaskie found among the perpetrators of Kumasi gun crime (2008a: 447), Sakawa boys defend their practice as taking something they deserve; a revenge against past and contemporary global and national inequalities: 'Kof-I: Dem say we do Sakawa cos we is greedy. Its all lies, lies like the white man from de times of de colonie [*sic*]. It is because r [*sic*] politicians is greedy and the world is greedy. Ther [*sic*] aint no chances for a black Ghanaian so we have to make dem!' (extract from Facebook conversation, 15 March 2010). Peter Geschiere suggests that witchcraft rumours bridge the global and local, with acute inequalities in a world of many desires accessible by few (1997: 81). As Bruno Latour stated, 'there are continuous paths that lead from the local to the global . . . only so long as the branches are paid for' (1993: 117, cited in Shaw 1997: 869). Sakawa offers access to these restricted 'branches' and their associated wealth and power, illicitly entering them via Ethernet cables to commit occult fraud. With many being refused visas or lacking the funds for travel, the internet allows access to places they cannot enter.

However, although Sakawa may transcend the frustration of global inequalities, the moral price is high. Whether it's the sacrifice of a local child or the reputation and prospects of the entire nation, Sakawa money is gained at the expense of others (Shaw 1997: 869). Sakawa boys' manipulation of the internet has strong parallels with long-standing occult fears that raise a number of ambivalences surrounding the internet's potential powers. Historical continuities claim neither origin nor explanation, but acknowledge that Sakawa witchcraft beliefs did not simply appear, as Sakawa combines long-standing beliefs with new powers of the internet (Shaw 2002: 11).

Manipulating the Power of the Internet

Many ethnographies have described witches' abilities to travel large distances in mysteriously little time. Rosalind Shaw's Sierra Leonean ethnography recounts rumours of witches flying to London and back within an hour (ibid.: 857). Similarly, the internet allows Sakawa boys to manipulate its compression of time and space, transcending national boundaries and engaging in worldwide spiritual travel. Shaw also describes how the Temne of Sierra Leone believe in an invisible, global space full of illicit wealth known as the 'Place of Witches': 'Although its presence may be recognized everywhere, its incompatibility with moral personhood and community is registered both in its intangibility and in its tropes of perverted and predatory consumption: the agency of the witch is necessary for the pursuit of one's desires within it' (ibid.: 857). Similarly, Ghanaians discuss the moral sacrifice entailed when a Sakawa boy makes an occult entrance into the intangible global space of the internet and extracts wealth by preying on foreign victims.

Sakawa boy's transcendence of national boundaries also has parallels with the Sakrobundi anti-witchcraft movement that originated in the Ivory Coast and spread to Ghana during the nineteenth century (McCaskie 2005). Colonial partitions and restrictions had created a tough economic climate, which promoted a smuggling trade that monetized social relationships and ultimately triggered a 'witchcraft craze', a craze that Sakrobundi offered protection from (McCaskie 2005:14). Like Sakawa, nineteenth-century witches gained their wealth at the expense of others, transcending the restricted economic context imposed by national borders whilst others struggled.

The powers Sakrobundi offered protection from were perceived of as 'alien' and, having entered from 'outside', were therefore especially dangerous (ibid.: 6). Sakawa power is also gained via drawing alien powers and wealth from the worldwide web using a technique that is believed to have spread from outside, i.e. Nigeria. Witchcraft and occult beliefs often involve the appropriation of foreign powers (Apter 1993). Sakawa is recognized for it's extraordinary, almost 'recharging' power to existing beliefs in illegal occult activities such as blood money or 'sika duro'. Sakawa rumours are the latest Ghanaian example of witchcraft and occult beliefs being perpetuated and reinterpreted in social space (Robertson 1996: 603):

Alice: Can you remember people doing sika duro (blood money) before, when you where younger?

Akosua: Oh yes, there has been sika duro for a long time. When I was schooling we heard of a girl on Aburi who was murdered for

	this blood money. The men, they took her organs and her sex and they used it so they could be rich and powerful.
Alice:	And now with Sakawa, is it different to sika duro or is it the same?
Akosua:	These witches, they are always getting more powerful and finding new tricks. They will trick you, you have to be sharp. This Sakawa, it is getting too popular . . . their spirit will enter the computers for the ju-ju. It comes from Nigeria; they are learning their tricks and getting more powerful. (Interview with Akosua aged 65, Akropong, Ghana, 1 August 2009)

Sakawa boy's power, as described by Akosua and many others, is a particular form of occult agency that is distributed via the internet. Anthropologists have commented on the often 'dividual' nature of witchcraft personhood, with witches having the power to be in multiple places at once; dispersing their agency via bodily transcendence whilst appearing physically present (Geschiere 1997: 127; Strathern 1988; Shaw 1997). Similarly, Sakawa boys are rumoured to appear as normal, sat in front of a monitor whilst their spirit possesses multiple foreign fraud targets. Sakawa movies such as *Sakawa Boys 2: Mallam Issa Kawa* (dir. Socrate Safo, 2009) often conclude with the demise of the Sakawa boy by his seeming physical presence yet spiritual death.

West African witches are also commonly believed to have extra organs such as eyes and stomachs that are used to entrap or consume their victims. Sakawa boys are similarly able to extend their bodies and consumptive powers via the internet – not via extraordinary physical traits but with numerous internet identities. Sakawa boys often have fake profiles on popular websites such as Facebook and Hotmail as well as on Western online dating communities. These profiles feature fake photos and thus allow the Sakawa boy to have multiple cyber bodies. Many Sakawa boys even switch gender using attractive female profile pictures to entrap fraud victims via online friendships or romance. It is therefore not surprising that many Ghanaians are wary of the potential powers that the internet offers.

Sakawa boy's cyber dispersal of agency, power and fame also has parallels with Malinowski's work on Melanesian Kula shell exchange systems (1978). In Kula, exchange magic is used to control the consciousness of the other Trobriand, persuading them to give up their valuable Kula shell. Similarly, Sakawa boys possess their foreign fraud target, persuading them to transfer money. Kula also extend and create personhood as Melanesians travel to multiple islands in systems of exchange, increasing their fame (ibid.: 51,75). However, rather than extending themselves through travelling by sea to other destinations, Sakawa boys' agency is extended through the internet. Daniel Miller has similarly drawn parallels between the internet and Kula in his work on Trinidadian social networking websites. Miller interprets these websites as aesthetically 'trapping' and able to 'influence the mind of potential exchange partners', promoting exchange with distant places and displaying a positive Trinidadian identity (2001: 137–88). Likewise, Sakawa boys use the internet to 'trap' their fraud victims, influencing their victims' minds via occult powers and fake profile pages. However, as discussed, unlike Miller's account of positive Trinidadian internet fame, the national fame spread by Sakawa boys is negative since it threatens Ghana's international reputation. Instead, Ghanaians discuss Sakawa boy's global visibility to give moral force to concerns surrounding an immoral way for Ghanaians to behave.

These dividual notions of witchcraft personhood, visibility and fame also raise interesting parallels with ideas of the 'Big Man', an idiom of power that has historical and contemporary value in Ghanaian society (Miescher 2005). Men of power are not 'big' individually but acquire their power via the division of their personhood through patron-client networks (Miescher 2005; Nugent 1995). The internet provides a new way to be dividual, which could lead to legitimate power and success for both individuals and their communities via online business and trade. However, Sakawa boys negatively manipulate this 'dividuality' with their internet fraud. Rather than capitalism and modernity creating a contested dichotomy between traditional obligations versus capitalist individualism (Geschiere 1997: 154), maybe Ghanaians are particularly wary of the internet because of increasing possibilities to be dividual and therefore powerful, but in a socially unacceptable and immoral way.

Ambivalence towards the potential positive powers of the internet is evident in Ghanaian concerns and desires surrounding internet cafes. Internet cafes are increasingly prominent throughout Ghana and are particularly popular among the youth. As a place for the young, they had become places of mystique and speculation for older informants. Often open until late, they were rumoured to sites for young people to practice their night-time Sakawa rituals.

However, the internet as an emblem of modernity and progress was commonly held alongside Sakawa fears among both young and old, with many informants articulating pride at the recent increase in local internet cafes. Among young people, cafes remained popular and local owners did not notice any decline in business. Although customers acknowledged potential Sakawa accusations, visiting the cafes was fundamentally a fashionable activity for local young people and was associated with being 'modern'.

This ambivalence illustrates that rumours of Sakawa boys' immoral manipulation of the internet do not mean that the internet should not be used positively. Witchcraft rumours surrounding modern idioms such as the internet are not a resistance to change (Geschiere 2007: 223). As Shaw's informant articulates, technology has immense potential within Africa: 'If we [Temne people] put such science to good use, what a great continent Africa would be' (2002: 210). As with Ghanaian's condemnation of Sakawa, many Temne people resent witch's appropriation of new technologies as selfish, blocking their opportunities to develop. However, occult power is also ambiguous, with the Temne not necessarily negating witchcraft's role in positive technological development. Shaw discussed Sierra Leonean attitudes to Western technologies as a mix of marvel and horror, with many believing them to be created by European witchcraft (1997: 860). Occult power is ubiquitous in African life; it is the way this power is used, rather that its existence, that is cause for concern. The manipulation of technology can bring success and witchcraft could be harnessed to reach this goal. Perhaps Ghanaian's critique of Sakawa is not an adamant condemnation of supernatural use of the internet, but of immoral supernatural use.

The intimate relationship between the internet and cultural archetypes surrounding the occult and ideas of power thus places Sakawa boys in a worryingly liminal and powerful space between the conceptual and physical borders of 'Moral' and 'Immoral' and 'Us' and 'Them', whilst simultaneously offering potential and danger. Many of these Sakawa ambivalences have been inserted into Pentecostal discourses as Ghanaians negotiate ideas of moral progress.

The Church and Sakawa: Contained or Sustained?

The church has taken an active role in assisting Ghana's 'moral progress', and pastors promote their responsibility to halt the Sakawa 'epidemic' via prayer,[13] as well as practically, by educating the youth about both the dangers and positive uses of the internet.[14] The church is a massive social and political force in Ghana and throughout West Africa (Gifford 1998; Marshall 1993; Meyer 1999). Ghanaian towns have multiple churches and denominations, from urban megachurches to those that pop up and down in backrooms or yards. Many have travelling pastors, websites and newspapers and are simultaneously part of a local and global Pentecostal community (Meyer 2002: 68). Like nationalism, African Christianity was predicted to eventually decline as an alien colonial relic. Yet since the 1970s there has been a Pentecostal 'second wave' (Gifford 2004) and many have noted the intimate contemporary Ghanaian relationship between Christianity and nationalism (Gifford 1998; Marshall 1993).

Christian discourses bridge the national and the local by involving both politicians and citizens (McCaskie 2008b; Marshall 1993; Meyer 1995). A Christian condemnation of Sakawa further inserts it into a national discourse where 'not Christian behaviour' and 'not Ghanaian behaviour' can become synonymous. Sakawa is one of many contemporary ambivalences that are negotiated by the church on a national scale. Tom McCaskie's ethnography on Ghanaian attitudes towards the recent oil discovery highlights the intimate relationship between Christianity, politics and 'progress' (2008b). Oil is inserted into Pentecostal discourses, with its potential prosperity often perceived as a gift from God but simultaneously a potential Satan (ibid.: 323). Politicians bickered over credit for the discovery, which would legitimate their power to bring prosperity to Ghana, whilst simultaneously inferring their religious right to rule (ibid.: 324–25). Often, national successes are Christian successes and national problems are Christian concerns,[15] and large churches promote their moral responsibility to halt Sakawa on a national level:

> We, as a nation, must continually praise God for His abundant blessings including the resounding victory of the 2009 Under 20 World Cup tournament, the visit of President Obama, commercial quantities of oil found, successful general elections . . . making Ghana one of the shinning [*sic*] stars on the African continent.
>
> While recounting and celebrating the goodness of the Lord, all serious-minded and well-meaning believers must be concerned about the total breakdown of our social fabric . . . such as Sakawa . . . armed robbery, bribery . . . just to mention a few. (Ansa 2009)

Pentecostal discourses are therefore often used to negotiate ambiguous issues; whether it be the oil discovery or Sakawa, Christianity can insert it into a good-versus-evil dichotomy. This dichotomy, and its diabolization of 'traditional beliefs' such as witchcraft, is often cited as a reason for Ghanaian Pentecostalism's popularity (Meyer 1999). The church offers witchcraft deliverance and protection, a space to confront a common fear (Moore and Saunders 2001: 16). Diane Ciekawy and Peter Geschiere (1998) have described African witchcraft as a force that need's 'containing', a role that Pentecostalism's good/evil dichotomy can provide. Witchcraft is acknowledged but is also domesticated and controlled via its condemnation as evil (ibid.).

Ghanaian witchcraft concerns often involve ambiguities surrounding consumption, with 'satanic money' featuring prominently within Pentecostal discourse (Meyer 1995). The problem is not money itself but the means of its acquisition, which must be moral and Christian (ibid.: 250). The church discusses the temptations and promises of modernity; promising prosperity if one has faith (Meyer 2002: 75). Ultimately Christianity is promoted as a moral route to progress and success (Marshall 1993: 234). The Christian imagination has therefore provided a space for traditional beliefs in evil spirits and forces to remain whilst negotiating contemporary concerns and ambiguities (Meyer 1999: 177).

However, the church may negotiate occult rumours but does not eliminate them; instead it spreads Sakawa rumours in its services and press. By simultaneously condemning Sakawa while revealing Sakawa practices, the church constantly reasserts itself as a purveyor of 'true knowledge' (Jordan-Smith 2001b; Meyer 1995). Power is asserted by revealing intimate details of witchcraft that may have previously been confined to secrecy or gossip (Meyer 1995). The church becomes a voice of authority on the occult, with many informants reiterating Sakawa rumours from church services during interviews.

Ultimately, just as Ghana seemingly needs Nigeria, against which to define its identity, Sakawa must be maintained to define what is not Christian behaviour. The church uses Sakawa rumours to reassert its identity and to bolster its popularity. Many large churches actively invest in the propagation of Sakawa rumours by commissioning large Sakawa posters that feature sensationalist Sakawa newspaper cuttings. Designed to deter and scare, these posters are an example of the church's role in maintaining witchcraft meanings in social space and promoting an alternative Christian route to prosperity. Filip de Boeck and Marie-Françoise Plissart's work in the Democratic Republic of Congo similarly concluded that the church is a 'crucial' contributor to the production and popularity of witchcraft in collective imagery (2006). The church has a vested interest in maintaining Sakawa fears; churches are big businesses and witchcraft fears promote attendance and therefore money collections (Meyer 2002: 69).

However, Ghanaians are not duped into attendance. Many are critically aware of the power and wealth of churches and rumours of corrupt pastors are common (Ciekawy and Geschiere 1998: 8; Meyer 2002; Jordan-Smith 2001a). Furthermore, reliance on witchcraft for identity and popularity means that the church treads a fine line. Daniel Jordan-Smith's work in Nigeria highlighted that although Pentecostal churches condemn witchcraft, they consequently place themselves 'precariously' close to it (2001b: 602). With the 'big business' of West African churches, pastors are often powerful and wealthy and are therefore vulnerable to witchcraft accusations themselves (Gifford 1994; Jordan-Smith 2001b; Meyer 1995). As in the Nigerian Owerri riots, whereby pastors were implicated in ritual murder related to 419 frauds (Jordan-Smith 2001a), some wayward Ghanaian pastors are rumoured to be Sakawa boys' spiritual masters.[16]

The church is therefore not just a condemner but a part of Sakawa rumours. By maintaining Sakawa rumours and inserting them into a good/evil dichotomy, the church reinforces a Christian and often 'Ghanaian' community. Although this can backfire with accusations directed at Sakawa pastors, Sakawa remains prevalent in Pentecostal discourse. During church services the congregation would often be instructed to pray for their health, their nation, the youth and the demise of Sakawa:

> And now let us pray. Pray for the health of our children, for the youth of our nation to be prosperous. Let our businesses flourish and our nation be a strong and Christian leader. Let us rid our society of these evil practices such as Sakawa, rid the devil from our lives and open our hearts to God! Only through prayer can we succeed! (Grace congregation, Ghana, Akropong, 9 August 2009)

Conclusion: The Ontogenic Dynamism of Sakawa Rumours

Beliefs in occult power that is gained at the expense of others are nothing new in Ghana and have a long history throughout Africa. Sakawa has been discussed in relation to previous witchcraft crazes such as Sakrobundi (McCaskie 2005) and has continuities with other West African ethnographies on witchcraft (Bastian 2001; Geschiere 1997; Jordan-Smith 2001b; Meyer 1999; Parish 2000; Shaw 2002). These comparisons highlight similar repulsive themes of shape-shifting, ritual murder, cannibalism, powerful snakes and the sacrifice of fertility (McCaskie 2005; Shaw 2002; Meyer 1995). However, Sakawa rumours are particularly interesting for their combination of these long-standing occult idioms with contemporary concerns on a national scale.

Sakawa rumours have been revealed as complex moral negotiations that engage a wide range of Ghanaians in the discussion of what is acceptable Ghanaian behaviour versus immoral Sakawa and its connotations with a stereotypical Nigeria. With increased reports of internet fraud, as well as the discovery of oil, Ghana has become increasingly similar to Nigeria and therefore potentially closer to its corrupt and immoral stereotypes. As is often the case with witchcraft rumours, Sakawa is being discussed at a time of social uncertainty and ambiguity (Douglas 2002: 27; Marwick 1982). Sakawa rumours are a process of national self-reflection, as Ghanaians note these similarities by defining Sakawa as being 'from Nigeria' and therefore recognizing its threat to their nation's international reputation and future prosperity.

Since Evans-Pritchard's famous monograph on Azande witchcraft, many anthropologists have highlighted occult beliefs as constitutive of a particular worldview (1976; Geschiere 1997). Sakawa rumours have been revealed as a particularly Ghanaian negotiation; however, this worldview is not limited to Ghana. It is literally a worldview, negotiating concerns for Ghana's global reputation and access to global resources. As James Ferguson stated in his 2006 book *Expectations of Modernity*: 'In Africa modernity has always been a matter not simply of past and present but also of up and down. The aspiration to modernity has been an aspiration to rise in the world in economic and political terms; to improve one's way of life, one's standing, one's place-in-the-world' (2006: 32). Sakawa boys ultimately betray their nation, threatening Ghana's international 'place-in-the world' and future prosperity.

These concerns propagate an epidemic of discussion, however the media rarely reports actual convictions and Sakawa confessions are rare. Although I spoke to Sakawa practitioners online, I never personally met a practicing Sakawa boy, nor encountered anyone who had. Ghana ranked seventh in 2008's US Internet Crime Complaint Top Ten List and rose to sixth in 2009, so internet fraud may be on the rise. However, Sakawa as a practice seems to be a predominately discursive phenomenon, or at least not on the scale that that rumours infer.

The discursive construction of an 'Other' has been discussed in William Arens's work on the worldwide pervasiveness of cannibalistic rumours that have been attributed to many, by many (1979). The Man-Eating Myth questions the reality of these accusations, instead highlighting their utilization in constructing a repulsive and often inferior 'Other' that maintains cultural boundaries (ibid.). Arens critiques anthropological accounts of the existence of cannibalism as mistaking rumours for fact. This mistake is motivated by anthropology's disciplinary reliance on alterity, which is situated in wider context of fascinations with a 'savage' or exotic 'Other' (ibid.). Sakawa rumours depict a number of socially grotesque images including cannibalism, but obviously Ghana's youth are not widespread ritualistic murderers. Following Arens, these popular images are discursive, involved in maintaining a Ghanaian cultural boundary.

Sakawa is particularly interesting for the change in scale of boundary maintenance. In Africa, witches are often believed to be jealous and selfish kin who betray their family values, destroying vitality and breaching the social boundary of the lineage. However, Sakawa rumours rarely point to kin; it is instead thought to be the practice of the abstractly defined 'Ghanaian youth'. Furthermore, Sakawa boys' occult manipulation of the internet allows them to have a global power.[17] It is not the boundaries of the lineage that are transgressed but the boundaries of the nation, and it is therefore Ghana that suffers (Ciekawy and Geschiere 1998). Ethnographies of witchcraft and the occult frequently interpret accusations as defining an 'Us' from 'Them', however the focus has often been on small-scale societies (Crawford 1967: 323; Douglas 2003: 132; Kluckholn 1944: 255; Lienhardt 1951: 310; Marwick 1982: 15; Mayer 1954: 68). With Sakawa these meanings have been extended to a national scale. Questions of 'who are we?', 'what do we stand for?' and 'what do we stand against?' have become redefined as Ghanaian concerns with contemporary relevance for the entire population. The Akan say 'it is the animal in your cloth that bites you'; Sakawa rumours illustrate that this 'cloth' has been extended to the nation and, consequently, so has the 'bite'. The internet provides young Ghanaian men with the potential to use occult powers to inflict misfortune on their entire nation.

Sandy Robertson has written of the extension of meaning, using an ontogenic approach to highlight the redefinition and transmission of meanings over time. Robertson values the life cycle of individuals in the transformation and accumulation of culture (1996: 591). As people grow, their understandings change, creating multiple meanings within a lifetime (ibid.: 598). Therefore as individuals' concerns and identity change over time so do their occult and witchcraft beliefs. This can be seen in 65-year-old Akosua's discussion of sika duro transforming into Sakawa as the internet and Ghana's global reputation become contemporary concerns.

However, rather than focus on the ontogeny of the body and changing understandings with age (ibid.), Sakawa rumour's dynamism has been discussed in relation to the growing nation state as self and society are redefined simultaneously (ibid.: 597). It is understandings of a Ghanaian identity that are changing, and Sakawa rumours reflect these concerns. A Ghanaian identity is not 'out there' to be engaged with, it is constituted by engagement. This identity is an amorphous mass, constantly transforming as Ghanaian's concerns and desires shift and change (ibid.). Sakawa rumours are maintaining and negotiating this identity and are sustained in social space because of their relevance to a large number of people who actively discuss Sakawa with their family and friends. For Sakawa rumours, this renegotiation is critically occurring as Ghana enters a new age as a successful African state, with expectations of its own, as well as from the global community.

As with many African countries and nations around the world, this is not the first time Ghanaians have had to negotiate who they are (Shaw 1997, 2002). From precolonial trade routes to the slave trade followed by colonialism, 'Ghana' has gone through many periods of change and 'culture contacts' (Gocking 2005; Hannerz 1992). As a result, borders, identities and fears have changed over time. Sakawa rumours highlight the ontogeny of

Ghanaian occult beliefs as meanings and concerns surrounding 'Ghana' grow and develop (Robertson 1996). Sakawa rumours negotiate Ghana's global reputation as an African success story as rumours of occult internet fraud threaten Ghanaian's hopes and expectations for the future. Ultimately, Sakawa rumours are involved in a contemporary and dynamic redefinition of Ghanaian identity and they will not be the last to do so.

Alice Armstrong completed her anthropology BSc (Hons) at University College London in 2010 and now works as a radio producer. This chapter is a revised version of her final undergraduate dissertation, which was supervised by Jerome Lewis.

Notes

1. 'Sakawa Reaction' Group 2009.
2. Quansah 2009.
3. For example: 'We must stop the youth doing this evil Sakawa. It is the work of the Devil!' (Wiafe and Botchway 2009). See also Amissah 2009.
4. Yamoah 2009.
5. http://www.ghanaweb.com/GhanaHomePage/features/artikel.php?ID=162565&comment=4774586#com.
6. 'Sakawa' could potentially be a non-Hausa speaker's approximation of *chakawa*, meaning 'to stab', similar to the word for spirit inflictions, *harbe*, 'to shoot'. Although the 'ch' sound is Hausa, the sound may be difficult for Ghanaian languages such as Twi where it is instead transformed into an 's' (M. Last, personal communication, 10 March 2010).
7. Post–oil boom, internet fraud has become increasingly popular among young Nigerian men who use a number of online techniques to scam foreign targets into transferring money into their accounts (Jordan-Smith 2007).
8. Newspapers were filled with national pride; see for example Tsen 2009.
9. Tawiah 2009. See also Asante 2009.
10. Online Sakawa discussion board comment by 'Lord Maxx' (Lord Maxx 2009).
11. The etymology of 'Sakawa' is not clear, and although it may derive from the Hausa *chakawa* (see note 6), it may also be related to *sakuwa*. This is a pidgin English word used in a 1960s Decca record entitled 'Stars from Zaire Volume 4,' by the Congolese Orchestre Bella Bella, and translates as 'you no resist' – i.e. greed (B. Sharpe, personal communication, 18 March 2010).
12. Charlanne Burke's (2000) study of Botswanan rumours of ritual murder known as Dipheko similarly illustrates tensions surrounding the embodied potential of educated youth in relation to national ideas of progress. However, Dipheko fundamentally differs from Sakawa as adults are the perpetrators against youth, whereas Sakawa involves critiques of youths' own behaviour.
13. From an online article by Reverend Patrick Kofi Amissah, St. Paul Methodist Church, Tema, Ghana: 'They have indeed become lovers of money and wicked people. This trend can only be stopped with God's intervention. . . . We need to pray and seek God's face to intervene so as to destroy the powers of the demonic authorities behind SAKAWA and other spiritually motivated frauds' (Amissah 2009).

14. 'Again, the church must find a very appropriate means of helping young people to use the internet and other modern technology positively. This can be through guided use by young people in the church's own café for this purpose, or through proper education for them to be vigilant' (Amissah 2009).
15. 'Church-State Must Fight Sakawa' was the headline of the *Christian Messenger* newspaper (Wiafe and Botchway 2009).
16. Sarpong 2009.
17. Other ethnographies have similarly noted contemporary witchcraft's often global scale, which incorporates long-standing occult beliefs and modern technologies (Ciekawy and Geschiere 1998; Geschiere 1997; Parish 2000).

References

Adinkrah, M. 2004. 'Witchcraft Accusations and Female Homicide in Contemporary Ghana.' *Violence Against Women* 4: 325–56.

Amissah, P.K. 2009. 'Internet Fraud from 419 to Sakawa – a Challenge for the Church.' *The Christian Sentinel*, 1 October. Available at http://www.thechristiansentinel.com/?p=872 (accessed 20 March 2009).

Anderson, B. 1991. *Imagined Communities: Reflections on the Origin and Spread of Nationalism*. London and New York: Verso.

Ansa, M. A. 2009. 'Is the Church Losing its Impact and Influence on Modern Society?' *Modern Ghana*, 8 December. Available at http://www.modernghana.com/print/253275/7/is-the-church-losing-its-impact-and-influence-on-m.html (accessed 2 April 2009).

Apter, A. 1993. 'Antiga Revisited: Yoruba Witchcraft and the Cocoa Economy.' In J. Comaroff and J. L. Comaroff (eds), *Modernity and its Malcontents: Ritual and Power in Postcolonial Africa*. Chicago and London: University of Chicago Press, pp. 111–29.

Arens, W. 1979. *The Man-Eating Myth: Anthropology and Anthropophagy*. New York: Oxford University Press.

Armah, K. 1974. *Ghana, Nkrumah's Legacy*. London: Collings.

Asante, D. 2009. 'Killer shot friend and bolted with GH¢96,200.' *Daily Guide*, 22 August, p. 3.

Author anon. 2010. 'Moulding Character of Our Youth.' *Graphic Ghana*, 8 March. Available at http://www.graphicghana.com/news/page.php?news=6694 (accessed 22 March 2010).

Auslander, M. 1993. 'Open the Wombs! The Symbolic Politics of Modern Ngoni Witch Finding' in J. Comaroff and J. L. Comaroff (eds), *Modernity and its Malcontents: Ritual and Power in Postcolonial Africa*. Chicago and London: University of Chicago Press, pp. 167–92.

Bastian, M. L. 2001. 'Vulture Men, Campus Cultists and Teenaged Witches: Modern Magics in Nigerian Popular Media.' In H. L. Moore and T. Sanders (eds), *Magical Interpretations, Material Realities: Modernity, Witchcraft and the Occult in Post-Colonial Africa*. London and New York: Routledge.

Bayart, J. F. 1993. *The State in Africa: the Politics of the Belly*. New York: Longman.

Berry-Hess, J. 2000. 'Imagining Architecture: the Structure of Nationalism in Accra, Ghana.' *Africa Today* 47(2): 35–48.
Birmingham, D. 1998. *Kwame Nkrumah: the Father of African Nationalism*. Athens: Ohio University Press.
Bleek, W. 1976. 'Witchcraft, Gossip and Death: a Social Drama.' *Man* (N.S.) 11(4): 526–41.
Burke, C. 2000. 'They Cut Segametsi into Parts: Ritual Murder, Youth and the Politics of Knowledge in Botswana.' *Anthropology Quarterly* 73(4): 204–14.
CIA World Factbook. 2010. Ghana. Available at http://www.theodora.com/wfbcurrent/ghana/ghana_people.html (accessed 1 March 2010).
Ciekawy, D., and P. Geschiere. 1998. 'Containing Witchcraft: Conflicting Scenarios in Postcolonial Africa.' *African Studies Review* 41(3): 1–14.
Comaroff, J., and J. L. Comaroff. 1993. 'Introduction.' In Comaroff and Comaroff (eds), *Modernity and its Malcontents: Ritual and power in Postcolonial Africa*. Chicago and London: University of Chicago Press, pp. xi–xxxvii.
———. 1999. 'Occult Economies and the Violence of Abstraction: Notes from the South African Postcolony.' *American Ethnologist* 26(2): 279–303.
Crawford, J. R. 1982 [1967]. 'The Consequences of Allegation.' In M. Marwick (ed.), *Witchcraft and Sorcery*, 2nd edition. London: Pelican Books, pp. 314–25.
Crick, M. 1982 [1976]. 'Recasting Witchcraft.' In M. Marwick (ed.), *Witchcraft and Sorcery*, 2nd edition. London: Pelican Books, pp. 343–64.
de Boeck, F., and M. F. Plissart. 2006. *Kinshasa: Tales of the Invisible City*. Antwerp: Ludion.
DiFonzo, N., and P. Bordia. 2007. 'Rumour, Gossip and Urban Legends.' *Diogenes* 5(4): 19–35.
Douglas, M. 1970. 'Introduction: Thirty Years after Witchcraft, Oracles and Magic.' In Douglas (ed.), *Witchcraft Confessions and Accusations*. London: Tavistock Publications, pp. xiii–xxxviii.
———. 2002. *Purity and Danger: An Analysis of Concept of Pollution and Taboo*. London and New York: Routledge Classics.
———. 2003. *Natural Symbols*. London & New York: Routledge Classics.
Ellis, S. 1989. 'Tuning into Pavement Radio.' *African Affairs* 88(352): 321–30.
Epstein, A. L. 1969. 'Gossip, Norms and Social Network.' In J. C. Mitchell (ed.), *Social Networks in Urban Situations: Analyses of Personal Relationships in Central African Towns*. Manchester: Manchester University Press, pp. 117–28.
Eriksen, T. H. 1993. *Ethnicity and Nationalism: Anthropological Perspectives*. London: Pluto Press.
Evans-Pritchard, E. E. 1976 [1937]. *Witchcraft, Oracles and Magic Amongst the Azande*. Oxford: Clarendon Press.
———. 1982 [1929]. 'Witchcraft Amongst the Azande.' In M. Marwick (ed.), *Witchcraft and Sorcery*, 2nd edition. London: Pelican Books, pp. 29–37.
Favret-Saada, J. 1980. *Deadly Words: Witchcraft in the Bocage*. New York: Cambridge University Press.
Ferguson, J. 2006. *Global Shadows: Africa in the Neoliberal World Order*. Durham, NC: Duke University Press.
Geschiere, P. 1997. *The Modernity of Witchcraft: Politics and the Occult in Post-Colonial Africa*. London: University Press of Virginia.

Gifford, P. 1998. *African Christianity: Its Public Role*. Bloomington: Indiana University Press.

———. 2004. *Ghana's New Christianity: Pentecostalism in a Globalizing African Economy*. London: C. Hurst & Co.

Gluckman, M. 1963. 'Gossip and Scandal.' *Current Anthropology* 4(3): 307–16.

Gocking, R. 2005. *The History of Ghana*. Westport, CT: Greenwood Publishing Group.

Grosby, S. 2005. *Nationalism: A Very Short Introduction*. Oxford: Oxford University Press.

Hannerz, U. 1987. 'The world in Creolization.' *Africa* 57(4): 546–59.

———. 1992. *Cultural Complexity: Studies in the Social Organization of Meaning*. New York: Columbia University Press.

Jordan-Smith, D. 2001a. 'Ritual Killing, 419 and Fast Wealth: Inequality and the Popular Imagination in Southeastern Nigeria.' *American Ethnologist* 28(4): 803–26.

———. 2001b. '"The Arrow of God": Pentecostalism, Inequality, and the Supernatural in South-Eastern Nigeria.' *Africa* 71(4): 587–613.

———. 2007. *A Culture of Corruption: Everyday Deception and Popular Discontent in Nigeria*. Princeton, NJ: Princeton University Press.

Kluckholn, C. 1962 [1944]. 'Navaho Witchcraft.' In M. Marwick (ed.), *Witchcraft and Sorcery*, 2nd edition. London: Pelican Books, pp. 246–62.

Konsom, F. 2009. 'Sakawa Could Sink Ghana says Lecturer.' *My Joy*, 20 May. Available at http://news.myjoyonline.com/technology/200905/30365.asp (accessed 25 October 2009).

Lewis, I. M. 1985. *Social Anthropology in Perspective: The Relevance of Social Anthropology*. Cambridge: Cambridge University Press.

Lienhardt, G. 1951. 'Some notions of witchcraft among the Dinka.' *Africa* 21: 303–18.

Lord Maxx. 2009. Comment 'Re: Nigerians are destroying Ghana'. 26 May, 01.20hrs. Available at http://www.ghanaweb.com/GhanaHomePage/features/artikel.php?ID=162565&comment=4774586-com (accessed 10 January 2010].

Malinowski, B. 1978. *Argonauts of the Western Pacific: An Account of Native Enterprise and Adventure in the Archipelagoes of Melanesian New Guinea*. London: Routledge.

Marshall, R. 1993. 'Power in the Name of Jesus: Social Transformation and Pentecostalism in Western Nigeria "Revisited".' In T. Ranger and O. Vaughn (eds), *Legitimacy and the State in Twentieth Century Africa*. Basingstoke: Macmillan.

Marwick, M. 1982 [1964]. 'Witchcraft as a social strain gauge.' In M. Marwick (ed.), *Witchcraft and Sorcery*, 2nd edition. London: Pelican Books, pp. 300–13.

———. 1982. 'Introduction.' In M. Marwick (ed.), *Witchcraft and Sorcery*, 2nd edition. London: Pelican Books, pp. 11–19.

Mayer, P. 1982 [1954]. 'Witches.' In M. Marwick (ed.), *Witchcraft and Sorcery*, 2nd edition. London: Pelican Books, pp. 54–70.

McCall, J. C. 2004. 'Juju and Justice at the Movies: Vigilantes in Nigerian Popular Videos.' *African Studies Review* 47(3): 51–67.

———. 2006. 'Nollywood Shows the Way.' *The Africa Report* 3: 120–24.

McCaskie, T. 2005. 'Sakrobundi and Aberewa: Sie Kwaku the Witch-Finder in the Akan World.' *Journal des Africanistes* 75(1): 163–207.

———. 2008a. 'Gun Culture in Kumasi.' *Africa* 78(3): 435–52.

———. 2008b. 'The United States, Ghana and Oil: Global and Local Perspectives.' *African Affairs* 107(428): 313–32.

Meyer, B. 1995. 'Delivered from the Powers of Darkness: Confessions of Satanic Riches in Christian Ghana.' *Africa* 65(2): 236–54.

———. 1999. *Translating the Devil: Religion and Modernity among the Ewe of Ghana.* Edinburgh: Edinburgh University Press.

———. 2002. 'Pentecostalism, prosperity and popular cinema in Ghana.' *Culture and Religion* 3(1): 67–87.

———. 2003. 'Ghanaian popular cinema and the magic in and of film.' In B. Meyer and P. Pels (eds), *Magic and Modernity: Interfaces of Revelation and Concealment.* Stanford, CA: Stanford University Press, pp. 200–23.

Miescher, S. 2005. *Making Men in Ghana.* Bloomington: Indiana University Press.

Miller, D. 2001. 'The Fame of Trinis: Websites as Traps.' In C. Pinney and N. Thomas (eds), *Beyond Aesthetics: Art and the Technologies of Enchantment.* London: Berg, pp. 137–57.

Moore, H. L, and T. Sanders. 2001. 'Magical Interpretations and Material Realities: An Introduction.' In Moore and Sanders (eds), *Magical Interpretations, Material Realities: Modernity, Witchcraft and the Occult in Post-colonial Africa.* London and New York: Routledge, pp. 1–27.

Nadel, S. F. 1952. 'Witchcraft in Four African Societies: an Essay in Comparison.' *American Anthropologist* 54: 18–29.

Nanaba, F. 2009. Comment: 'National Cake'. 26 May, 18.25hrs. Available at http://www.ghanaweb.com/GhanaHomePage/features/artikel.php?ID=162565&comment=4776955#com (accessed 10 January 2010).

Nugent, P. 1995. *Big Men, Small Boys and Politics in Ghana: Power Ideology and the Burden of History, 1982–1994.* London: Mansell.

Olivier de Sardan, J. P. 1992. 'Occultism and the Ethnographic "I": The Exoticizing of Magic from Durkheim to "Postmodern" Anthropology.' *Critique of Anthropology* 12(1): 5–25.

Oomen, T. K 1997. *Citizenship and National Identity: From Colonialism to Globalism.* New Delhi: Sage Publications.

Opoku, E. 2009. 'Another Sakawa Grabbed!' *Daily Guide,* 25 August, p. 1.

Owusu, D. 2009. 'Nigerian High Commissioner Cannot Deny Ghana Bashing.' *Ghana Web,* 19 September. Available at http://www.ghanaweb.com/GhanaHomePage/features/artikel.php?ID=168942 (accessed 10 March 2010).

Parish, J. 2000. 'From the Body to the Wallet: Conceptualizing Akan Witchcraft at Home and Abroad.' *The Journal of the Royal Anthropological Institute* 6(3): 487–500.

Quansah, M. 2009. 'Sakawa is Ruining Ghana.' *People and Places,* 22 July, p. 1.

Ranger, T. 1993. 'The Invention of Tradition in Colonial Africa.' In E. J. Hobsbawm and T. Ranger (eds), *The Invention of Tradition.* Cambridge: Cambridge University Press, pp. 211–63.

Robertson, A. F. 1996. 'The Development and Culture: Ontogeny and Culture.' *The Journal of the Royal Anthropological Institute* 2(4): 591–610.

Rowlands, M., and J. P. Warnier. 1988. 'Sorcery, Power and the Modern State in Cameroon.' *Man* (N.S.) 23(1): 118–32.
Sakawa Boys: Mallam Issa Kawa. 2009. Feature film. Directed by Socrate Safo. Kumasi: Ghana.
Sakawa Boys 2: Mallam Issa Kawa. 2009. Feature film. Directed by Socrate Safo. Kumasi: Ghana.
'Sakawa Reaction' Group. 2009. Facebook group, available at http://www.facebook.com/home.php?#!/group.php?gid=102575958216&ref=ts (accessed 23 February–10 April 2009)..
Sanders, T. 2003. 'Reconsidering Witchcraft: Postcolonial Africa and Analytic (Un) Certainties.' *American Anthropologist* 105(2): 338–52.
Sarpong, D. 2009. 'Top Pastors in Ghana Patronize Sakawa Boys.' *Today Newspaper*, 4 May. Available at http://www.theghanaianjournal.com/2009/05/04/top-pastors-in-ghana-patronize-sakawa-boys/ (accessed 11 December 2009).
Shaw, R. 1997. 'The production of witchcraft/witchcraft as production: Memory, modernity and the slave trade in Sierra Leone.' *American Ethnologist* 24(4): 856–76.
———. 2002. *Memories of the Slave Trade: Ritual and the Historical Imagination in Sierra Leone*. Chicago: University of Chicago Press.
Shibutani, T. 1966. *Improvised News: A Sociological Study of Rumor*. Indianapolis, IN: Bobbs-Merrill.
Stewart, P. J., and A. Strathern. 2004. *Witchcraft, Sorcery, Rumours and Gossip*. Cambridge: Cambridge University Press.
Strathern, M. 1988. *The Gender of the Gift*. London: University of California Press.
Taussig, M. 2003. 'Viscerality, Faith, and Skepticism: Another Theory of Magic.' In B. Meyer and P. Pels (eds), *Magic and Modernity: Interfaces of Revelation and Concealment*. Stanford, CA: Stanford University Press, pp. 272–306.
Tawiah, O. 2009. 'Ashanti Police Hunt for Armed Robbers.' *Daily Guide*, 4 August, p. 7.
Transparency International: The Global Coalition Against Corruption. 2008. *Corruption Perceptions Index 2008*. Available at http://www.transparency.org/news_room/in_focus/2008/cpi2008/cpi_2008_table (accessed 1 March 2010).
———. 2009. *Corruption Perceptions Index 2009*. Available at http://www.transparency.org/policy_research/surveys_indices/cpi/2009/cpi_2009_table (accessed1 March 2010).
Trouillot, M. 2001. 'The Anthropology of the State in the Age of Globalisation.' *Current Anthropology* 42(1): 125–38.
Tsen, K. 2009. 'HAIL GHANA THE NEW FACE OF AFRICA.' *Daily Graphic*, 13 July, p. 1.
US Internet Crime Complaint Top Ten List. 2008. Annual Report. Available at http://www.ic3.gov/media/annualreport/2008_ic3report.pdf (accessed 5 March 2010).
———. 2009. Annual Report. Available at http://www.ic3.gov/media/annualreport/2009_IC3Report.pdf (accessed 5 March 2010).

Wollaston, S. 2010. 'Blood and Oil and Panorama: Passports to Kill.' *The Guardian,* 30 March. Available at http://www.guardian.co.uk/tv-and-radio/2010/mar/30/blood-and-oil-panorama-passports-to-kill (accessed 1 April 2010).

Wendl, T. 2007. 'Wicked Villagers and the Mysteries of Reproduction: An Exploration of Horror Movies from Ghana and Nigeria.' *Postcolonial Text* 3(2): 1–21.

Wiafe, I., and F. Botchway. 2009. 'Church-state Must Fight Sakawa.' *Christian Messenger,* 17–31 July, p. 1.

Yamoah, E. 2009. 'Mob Battle Police over 'Sakawa' Boys.' *Daily Graphic,* 26 July, p. 5.

Chapter 5

To Heal the Body

The Body as Congregation among Post-Surgical Patients in Benin

Isabelle L. Lange

For his thirty-second birthday a few years ago, Alexandre invited friends and family to his friend Serge's house on a Sunday afternoon in the village where he grew up in southern Benin. Sixteen years previously, a small, crusty wound had appeared on his lip that would reveal itself years later to be cancerous. Despite various treatments in his hometown and in nearby Cotonou, the country's economic capital, the wound ultimately took over the majority of his face beneath his eyes. It had been a couple of years since he had undergone full eliminative and reconstructive surgery on board a travelling Christian hospital ship docked in Benin, and later in Ghana, and there had been an air of hope and promise in his dialogue and that of his friends. However, the cancer returned after surgery and resumed its steady course across his face, and on this occasion the language and mood surrounding his activities was decidedly sombre. Guests did not speak of his death, but neither was there any planning for a future; the woman he had had a crush on from afar in the past had not been replaced by a new one, and the ideas he had the previous year to start up a cybercafé were not brought up. As we sat around the table at lunch it became time for the formal speeches. The tone was serious, as people reflected publicly. 'Your job is to persevere, and to keep on living each day as you do, and our job is to keep praying for you, to keep the faith,' said Serge, a director at the school Alexandre once worked at, and his closest ally.

Notes for this chapter begin on page 112.

This chapter looks at the corporeality of an illness and the communities of individuals that band together to carry a sick person through their illness, while simultaneously strengthening a sense of communality along the way. Based on fieldwork in southern Benin surrounding the crew and patients of a Christian hospital ship, I explore the interpersonal and spiritual networks that formed as a base beyond the surgical and evangelical interventions of the ship that, with its arrival, joined a sort of congregation made up of supporters of those who are sick. The congregation is the social network that represents the social body. In the case of Alexandre, it is mobile, following along his life path. The ill body becomes the focus of attention and facilitates social cohesion. It is a call to assembly for those who choose to answer, beyond family and friends and the church, to a selective greater community that unites in order to assist an individual who is suffering, such that those who become marked by the illness are not only those who are 'saved' but also those who do the saving. By looking at the religious responses enacted through individuals, a patchwork is woven wherein it is not just the ill body that is taken care of but the body of the congregation and its faith. This chapter explores how going through illness and suffering can separate individuals from a group, and how the embodied act of hope and searching for ritual experiences can draw them back into meaningful experiences of mutual transformation, creating a cycle where faith is healing, and healing becomes faith.

Background Questions

The hospital ship, being part of a non-governmental organization dependent on private and corporate money, in its marketing campaign for funding and sponsors presents its patients to the Western world as being outcasts in society. Plagued by the stigmatizing nature of their illnesses, they are shunned by those who think they are cursed and brought the illness onto themselves. While things may be as the ship describes to some degree, I was sceptical of this claim when I read their materials before I went to the field, and indeed, once I started meeting patients off the ship in their communities, I was struck by how many people rallied around them. I started to think about not only how many people were in the patients' social networks, but also who exactly they were. I came to see that it was a particular mix of family members and community members, and was interested in exploring how these groups that surrounded the patients came about. Why did some people stand by those who were sick, when many other people turned their backs? Could it be as simple a question as the mystery and beauty of friendship, of altruism, of Christian Love – or was there more involved? And in any case, what was the fabric of these relationships? What bound them together? Not all family stayed with the ill person throughout their illness, but people active in their lives shifted over the passage of time. Those that were helping in the traditional sense were also those that were being helped along their own passage through life.

As a bridge between the encounters between land and ship, broadening the actors to include family and friends, and as an integral piece of people's personal transformations, this chapter proposes the idea of congregation. It presents the voices of some of the crewmembers, not only to illustrate their feelings and motivations for working in this healthcare setting but also to show how the resulting community is constituted through ritual acts of believing. The hospital ship is a point of departure for this exploration, as it creates a mobile community that travels, gives power to ideas through its outreach and healing and carries its medical work back into communities through the people treated. Then I will shift back to land – to the patients of the ship – to look at the elements of pain and suffering that factor into creating a lifeworld apart for a sick individual. By focusing on a prayer session that Alexandre is involved in, I shall also examine how this ritual act creates meaning and helps make sense of the world for the individuals who choose to be involved in his life.

A Sense of Self and Illness

By looking at the impact of ritual acts on a community, I want to explore the notion of faith – its struggled maintenance and its renewal – as healing, and healing, in all its incongruous, inconclusive ways, coming to constitute faith. 'Out of illness experiences emerge interpretations that see temporality and the body as integral to the process of ordinary living,' write João Biehl, Byron Good and Arthur Kleinman (2007: 31), highlighting the arena illness provides for an exploration of the body's contestation and restitution within contexts. I start by looking at one idea of what embodies a person's sense of self, and notions of congregation and the healing of sickness, both physical and social. This is in order to explore the concepts that articulate a person's creation of ways of being in the world, and elucidate the agency they have in influencing their lives and those of others, as a way of moving towards seeing one's personal contribution to a community.

The perception of an individual's ways of being in the world and his/her relationships to others is integral in identifying a personal sense of self. In writing and researching personhood, many contemporary authors reflect back on the work of Marcel Mauss (1985 [1938]) and Meyer Fortes (1987 [1945–49]). Central to Mauss's concept of the *personne morale* is the idea that personhood is created and affirmed through social relationships – as he or she negotiates the inner self with the outer person that is socially formed, and thereby becomes aware of him/herself. Fortes responds to this essay with ideas on the concept of the person gleaned from his fieldwork among the Tallensi of northern Ghana, and stresses that personhood is something achieved little by little throughout the life course through rituals and key events.[1] A sense of self, and thereby, one's personhood, is something constantly in flux and evolving, and on top of that, not even 'owned' by oneself. A person has as many selves as there are other people who have knowledge of that person and reflect on them themselves, observed William James (1950, cited in Jackson 1998). There is no controlling this image, as it is inextricably linked to the pictures and impressions of other members of one's group of interaction. It follows, then, that one can manipulate to some degree who one 'is' through the people one spends time with, and who one allows to reflect images of oneself back. A person's way of being in the world is tethered to how they are perceived by others.

The trend since the 1980s towards 'bringing bodies back in' (Frank 1990) to the study of personhood has gone against a long-standing tradition of taking a disembodied view of socialization. Subsequently, much of the movement in the work done on the body in anthropology has been towards a theory and practice of embodiment. Much work now challenges, and denies, the Cartesian split of dualism and seeks to explore the critical place of the body in society, culture and personality. However, Thomas Csordas (1994) called for a step back from mistaking the body for personhood, in response to a trend he noticed occurring when anthropologists were, in his mind, writing about the body in culture without any sense of 'bodiliness' – that which makes bodies more than simply things and gives them 'intentionality and intersubjectivity' (1994: 4). Research and analysis done this way, he says, miss key opportunities to add sentience and sensibility to how we view the person. Bodies are the site of emotions and of the personal, even in their public representations. Its functions and abilities are seen as symbols of integrity, and the consequent reactions – rejection, disgust, fear – become a manifestation of social control through bodily control (Douglas 1970). Where does the body end and the social and spiritual world pick up?

This social control and oppression can result in someone being cast out, stigmatized and denied access to a group, based on a physical deformity. In the case of Alexandre, for example, his bodily expression of a grave illness was considered by some to be a physical manifestation of the (negative) energies affecting him. It was also a complex centre of contestation surrounding the source of affliction and the recourses to heal in this landscape where many hold to a Christian belief to counter the feared capacities of other local religions. This idea of stigma crosses over with the notion of the outsider as conceived by Hans Becker (1963). The outsider is a person who fits a social definition of deviance that locates the outsider not so much outside but very much inside of a particular society and its beliefs. Myths, fears and misunderstandings grow up around stigmatized and 'outsidered' bodies (Murphy 1990), which create barriers that must be overcome in order for the person with a stigmatized body to occupy a normalized societal role.[2] The disabled person suffers a 'contamination of identity' as they are not seen as anything beyond the reach of their bodies – whether that is interpreted positively or negatively.

Thinking about what is happening at this site of social regeneration and creation of congregation evokes John Janzen's (1982) research on the repositioning of individuals in a community after a change in health status. What in some places would be perceived of as an illness is interpreted in his fieldsite in Central Africa to be practically of honorary status, in that men with scrotal hernia are instead elevated to being of their own class, a process affirmed through a ritual that marks the participants' new role in society. This particular condition is not perceived as a negative affliction, whereas in other contexts it might be seen as an impairment and stigmatized. According to Janzen's analysis, this system at the same time provides a kind of therapy, in order to return power to individuals – by reintegrating them as part of the group – who might otherwise have lost some of their standing and possibilities. In order for this to work, the ritual for improving the condition must fit to the meaning system of the given culture. As we will see in the following sections, in this setting a similar phenomenon takes place that is beneficial to those who choose to be involved. Getting well involves the adoption of social and other routinized processes by which wellness can be distinguished from illness. A 'well' body may not be free of disease, but can at least have the social networks and rituals that impart on it meanings beyond those that are burdensome and negative.

In describing the processes illustrated in this chapter I use the term 'congregation' to refer to a body of individuals who come together to worship – that is, to seek worth through ritual and devotion. A deciding characteristic of this group is not the unified physical meeting of individuals that bonds people, but the social networks, intentions and repetition that act as points of cohesion. In this sense, it is a conscious, voluntary, self-selected body of people that are bound together through the locus of a physical body that calls to be supported and healed. Throughout this chapter, we will see various congregations form at the face of the diseased body: the social network of the patient and the ship (crewed by a team of individuals) coming over water to offer hope and healing. The intentions here are to restore a social order, to find meaning for oneself through the suffering and to alleviate the suffering of another (sick) individual and oneself, all through which a congregation is drawn together and reorbited.

Having a sense of self represents one way of being in the world. When things become disrupted (for the sick) and distorted (for the community), processes need to happen in order to transform the suffering and interference into an adjusted world order. 'Sickness is not just an isolated event, nor an unfortunate brush with nature', write Nancy Scheper-Hughes and Margaret Lock, 'it is a form of communication . . . through which nature, society and culture speak simultaneously' (1987: 31). Just as a collective can form its identity through the stigmatization of a group or the rejection of a sick person and a disease by adhering to a set of agreed-upon beliefs, superstitions and knowledge, others can instead embrace and nurture, and likewise contribute to their sense of being in the world on the back of this person or situation. Individual transformations are conscious and agentive: people transform themselves to deal with the constraints presented.

Placing the work that the hospital ship does into the movement it contributes to, as further discussed in the next section, lends another facet to this manifestation of congregation. In worlds that separate further from one another as they become more complex, often the acts of coping and adapting involve a splicing off of oneself into an assortment of groups, belongings, geographies and meanings. Development approaches and medical work went through an incarnation of being solution-oriented and tackling pieces of the whole instead of the whole itself. Kleinman writes: 'Institutional responses tend to fragment these problems into differentiated, smaller pieces which then become the subject of highly particularized technical policies and programs, increasingly ones that last for short periods of time and then are replaced by yet others which further rearrange and fracture these problems' (1999: 392). In this sense, the hospital ship (in common with other faith-based organizations) counters institutional measures that fracture people, communities and their problems by employing faith and a vision of religion to unify these aspects of their lives. Certain individuals in the patients' home networks back on land pick up from here and offer a continuum for this; those that participate counter the rupture threatened by those who do not. This congregation affords an opportunity to piece it back together.

The Ship as a Site of Congregation

With its on-land headquarters based in Texas, a crew of about 350 people run the ship along its course primarily along the coast of West Africa, docking in ports for about four to seven months and carrying out a combination of community and medical projects, the centre of which are cost-free surgeries for the local population. In Benin, the organization works, at least officially, in collaboration with the Ministry of Health to establish goals that both parties are interested in meeting. At the time of fieldwork (2004–2009), the lack of a functioning medical insurance system and low wages made the medical procedures the hospital ship offered an otherwise unattainable dream for the majority of the Beninois population who might have needed them and been able to locate a local provider.

The evangelical aims of this non-governmental hospital ship are clear, even if the medical assistance they offer comes with no strings attached, so to speak. Historically, the original way 'into' a culture for religious missionaries was through medicine – nothing else was as impressive as healing a body. Healing involved the body, mind and spirit and could be a means of negotiating the boundaries and conflicts caused by the colonial encounter (Comaroff 1997: 323). Purification campaigns were at the forefront of efforts to proselytize and cure the body, and consequently the soul (Masquelier 2005: 6).

This organization is part of a wide network of faith-based initiatives in development and medical aid. The non-governmental organization (NGO) taxonomy includes secular and faith-based organizations (FBOs). They both maintain a strong presence in the region, and this classification becomes meaningful not only in distinguishing between secular and non-secular organizations, but also in examining the differing characteristics of non-secular organizations themselves (Thaut 2009). As NGOs have been on the rise, the number of FBOs active in the international development field have also increased in the same time period, particularly since the 1990s. A study of US aid organizations between 1939 and 2004 working internationally showed that 'evangelical organizations accounted for 33% of all relief and development agencies and 48% of the total number of religious humanitarian agencies' (McCleary and Barro 2006).

Distinguishing between FBOs and secular organizations is not always straightforward. Unlike the narratives of early missionary work, the 'faith' in FBOs is not necessarily about impacting (or converting) the spiritual lives of recipient populations, but instead can refer to the organization being driven by its faith to carry out its mission. As Richard Falk (2001) notes, faith also contributes a duty-oriented aspect to organizations that, while perhaps encompassing some of the same values of non-FBOs such as taking a rights-based approach, comes from a place where the underlying origins of the notions of justice, reconciliation and transformation differ.

Many in the secular development world tend to be sceptical of them, but in Benin, FBOs fit into the local schema quite well (as will be seen later through stories of interactions between the ship and local communities). When overt religious agendas are not pushed (and also in some cases precisely because they are) they are able to find a home and succeed in these areas, at least from a standpoint of shared values. FBOs have a reputation for taking the whole person into account (not just their economic unit) in terms of their project, and are characterized by a vast network of individuals, congregations and other NGOs (Berger 2003). One aspect that can distinguish FBOs from other NGOs is that the former will attempt a more holistic approach to their aid, which follows along Erica Bornstein's observation that in 'African Christian faith' the 'realms of the spiritual and material cannot be easily separated: development is both spiritual and material' (2005: 49).

Within the social tradition of linking suffering and charity (Redfield and Bornstein 2011), there has historically been a strong link between Christianity and humanitarianism (Fassin 2012). The ship's mission is to bring 'hope and healing to the world's forgotten poor'.[3] Internal and public ship documents expound on this purpose: 'The nature of our work is essentially redemptive and restorative. In effect, our work is to walk alongside the forgotten poor, facilitating and participating in their growth as well as our own.'[4]

Evangelism[5] is important to those who come on board, but the religious dimension is just as important a motivation for internal ship operations as it is for activities off the ship. Emboldened and inspired by their religious belief to 'do good unto others', the act of carrying out this aim is considered by many on board to be more important than the aim of actual conversion of non-believers to Christianity. This was evident in that actively demonstrating one's Christianity *on* board was important as much as acting Christian was encouraged both on and *off* ship, especially during outreach into different communities. While patients receiving medical care on board could stay with little evangelical focus addressed in their direction, crew could not. As Matt Tomlinson and Mathew Engelke have stated, 'notions of intentionality often become central to Christian conceptions of "faith"' (2006: 14). The intention here is not simply to convert others, but to convert oneself – to transform oneself into a better Christian and to have an evolving relationship with God.[6] What I wish to describe about this community goes beyond simply notions of evangelism, but extends into the personal messages of communion that are repeatedly enacted through living and believing in the potential of this congregation.

Volunteers recognize that setting a good example of their belief is one way of inspiring others to follow suit. The FBO's media material made available to the press, funders and also to crew repeatedly highlights instances in which a non-Christian surgical patient was so overwhelmed by the generosity and affection he received from the attending medical staff that he decided to convert and allow Jesus into his heart. While this can be one product of the volunteers' missionary work, the other is that they themselves benefit from their own efforts, which was shown in a conversation I had with a retired British man during his fourth stay on the ship. His job working with the onboard-services department kept him away from regular interaction with patients, but he volunteered at the medical screenings, offering water and answering the questions of people waiting in the queue sometimes for hours. He mentioned meeting a young man with a large facial tumour, larger than any he had seen before, which he described in detail to me. He said he was repulsed and fascinated at the same time, and that he was overcome with an 'internal dilemma' when he sat with the man: 'I touched him – I don't think anyone had touched him in a long, long time. And I sat with him and looked him in the eye. . . . But, you know, I think I did it for myself, as much as for him. I was curious. I think I wanted to prove to myself that I could reach out to this extremely sick man, that I wasn't scared of him.' This exchange came at the end of a conversation about his reasons for coming to Africa to help the poor and sick instead of helping disadvantaged people in his hometown in England. Although he had been involved in community service projects at home, this approach, he said, interested him at this point in his life and felt like a greater challenge.

Personal expectations for transformation and change are a driving force in this community of individuals. Volunteers are focused not only on the world outside the ship (by improving health and/or enabling spiritual conversions) but also on themselves. Near the beginning of her time serving on board, a crewmember posted an entry on her online journal, entitled 'And God saw that it was good', in which she anticipates the transformation she will undergo once the activities involving patients get under way, the first being the screening for new patients scheduled shortly after the ship's arrival (to select those who will receive medical treatment during the coming months):

> I'm trying to mentally prepare myself for tomorrow. The person writing this post is not going to exist tomorrow. The people I meet and film are going to change me forever. Tomorrow is a day I have been waiting for for eight years. That is how long I have wanted to serve with Mercy Ships.
>
> I woke up this morning, late I might add, and as I tried to mentally prepare myself for tomorrow the book of Genesis came to my mind. In the first chapter

> one phrase is repeated, 'And God saw that it was good'. The word 'saw' is repeated seven times.
>
> Tomorrow I want to remember this when I am looking at people who are outcasts of their society. God sees them! He saw them when he knit them together in there [sic] mother's womb. He sees their struggles and calls them to Himself. I want to see them the way He does.[7]

This entry is notable for several reasons. First, she emphasizes the visual aspect of the sick, the corporeality of illness and the potential impact of the act of witnessing. Second, she is willing a transformation for herself into being and to initiate it she reaches out to the patients, seeking a way to come into contact with hurtful, difficult pieces of life (by filming the screening) and through these images and experiences also be healed by them. She seeks out that which, ultimately, she believes will help her evolve in her relationships and become who she wants to be. Being there, and being involved, will irreversibly change her way of being in the world. Furthermore, she initiates this process by writing about it and calling upon her community of readers online to witness her transformation and also participate in it themselves. The comments in response to this and other blog entries echo her calls and affirm her path of seeking.

Screening days are exciting, long and often overwhelming, requiring an immense amount of organization in order to get through the hundreds, often thousands, of individuals who have lined up in order to be seen by the ship's medical staff in the hope of being selected for treatment. Not everyone waiting makes it into the reception hall, but those who do are generally examined and determined by the team to be eligible for medical treatment or not. If they are not, the individual and any accompanying family members are asked if they would like to be escorted to a separate area on the grounds where they are met by ship volunteers and translators to guide them in prayer. This prayer option features strongly in the dialogue and discourse of the volunteers. Crewmembers working at the screening are confronted with the visual sight of a queue of people who have travelled from as far as hundreds of miles away, who have been suffering without medical care for many years and who have in some cases been sleeping outside to be sure to get a space at the screening. The massive line numbering in the thousands and the charged emotions can be overwhelming for the crew, knowing that they will have to turn the majority away.

One evening in the ward, while spending time with Alexandre, I happened to overlap with Betty, a crewmember whose role involved providing emotional support to patients both on and off the ship. We chatted about some of the characters staying on the ward and about the challenges of her job. As we turned to talk of God I recounted that an academic theoretical take on religion and illness is that religion's role is to change the meaning of the illness for the sufferer, and I asked her what she thought of that. Betty laughed and responded that she simply wanted to make people feel important and loved; she did not know what else religion would do for them, but that is what she liked to do with it. In this case, for Betty the formal, concrete outcome was not the goal, but the enactment and re-enactment of her symbols of meaning. How she wanted her patients to feel reflected back onto how she made sense of her world.

This was demonstrated in a primarily Muslim (and 'animist', according to crewmembers) village in Sierra Leone, about an hour and a half from the capital, Freetown, where the ship was docked. Volunteers packed into two Land Rovers had driven out of the city to host an evening viewing of *The Jesus Film*[8] once the sun set enough to be able to watch it on a makeshift screen rigged up on the back of one of the 4×4s. Outreach activities such as this were central to the onboard ethos (beyond their evangelical function); for many crewmembers they were the only opportunities to have direct religious interactions with local communities. Crew who did not themselves regularly volunteer heard about these experiences as stories were fed back through both informal and formal lines of communication on board.

The evening's film viewing concluded with a local pastor preaching to the assembled crowd of villagers, calling all who wanted to cross over to God to stand up and step across a line he drew in the sand. A few people got up, crossed over and lingered on the other side. Crewmembers mingled among the audience while others packed up the technology, and amidst this a parent carried his blind child to some volunteers. Almost silently, in a fluid motion, three volunteers gathered around the child and his parent, motioned to another crewmember to join them and began to pray in unison for the life of the child, asking for God to bless him. They sat in the centre of the circle and the volunteers joined arms, guided by a path of ritual prayer that did not need to be articulated or discussed among them.

'That was great, wasn't it?' exclaimed a Swedish woman exhausted and bouncing around in the back of the truck on the way back to the main road to Libreville and to the ship. 'We were able to do so much Good.' I asked one of them later to talk more about the sentiment of 'Good', and she explained that because of the negative energy in the village they had had to do what they could to respond to it. In order to be stronger than the 'evil' in the village, they could together call upon God to make things right. Having a code that provided guidance on how to respond not only helped them individually come through this sad encounter but also let them share it with strangers, with the intention of offering comfort and feelings of hope to them.

Messages and enactments of faith and kindness take different forms and labels across contexts. Julie Livingston explores the Setswana concept of *botho,* a term recently entered into the public discourse of national identity that refers to a 'Tswana ethic of humanness, which acknowledges that one's actions affect others' (2008). The *botho* concept has been around in the local context for a long time, and involves compassion and a sense of 'humanity' that are challenged by the revulsion experienced when faced with bodily aesthetics that do not conform to the norm. *Botho* calls for people to be kind, considerate and accepting of one another. However, physical disabilities and illnesses – those that illicit reactions of disgust from others – test the boundaries of those who normally practice *botho* but do not or cannot any more. While I did not come across a similar term in Benin during fieldwork, I found this idea useful when looking at the FBO's work in the region and the changing relationships patients had with their caregivers. Describing something parallel to this, the term 'Christian Love' frequently came up. Some locals referred to the crew of the ship as 'angels sent from heaven' because of the work they did with the sick, but I am looking at this concept as being as much about the impact on those embodying the spirit of *botho* as those who are the beneficiaries of it.

Ignoring the fear that such a facial deformity could provoke is one thing, but it can be equally as difficult to ignore the aesthetic facts of illness, in this case facial disfigurement. Often there is exposed, rotting flesh that gives off a pungent smell, and the tumours that grow out of the mouth can make eating or drinking very messy affairs, with drool and masticated food getting stuck in view. I mention these things because of the tension and discomfort I witnessed during the rare occasions of eating in groups with some patients before their surgery (or, in Alexandre's case, after). These occurrences, which could be viewed as 'matter out of place' (Douglas 1966), can be interpreted by others as being evidence of a lack of virtue, a failure to have played by the rules. In Julia Kristeva's (1982) study of the abject, it is these people who can be excluded by individuals in order to create their identities, and through these acts of segregation on a larger scale construct the values of a society. In the context of this research, it is those close to God – pastors and medical staff as well as all other believers – that are called upon to exercise Christian love and practice the so-termed *botho,* even when they are placed in positions that challenge them. Many of the crewmembers try to visit patients that they have met on board, or other locals, at home, and while part of this is for their own recreation, in addition it is to demonstrate acceptance where they expect there to be little given by the local community. In particular, there is an outreach arm in which two nurses visit a small number of individuals who have been determined to be too ill to be operated on. These visits into the community by the crewmembers (foreigners) are meant to set an example of generosity for the patients' neighbours, in addition to offering immediate support where needed.

The ship as a body works to heal both its patients as well as its volunteers working on board, who seek out a transformation and have the desire to transcend their individual limits collectively. Unified intention generates social cohesion. The ship acts as a point to tie together the individuals of singular intention but diverse origins, led not by a charismatic leader so often highlighted as being the illuminators of group movements (see many, but Csordas 1994 and Harding 2000 as examples), but by the product itself, acting as a healer for those who heal. Here it is not any one individual who is charismatic and leading a revolution in healing, but it is the social cohesion and intent that creates the charisma. The charisma, in a sense, is the ship itself, the embodiment of unified intentions and recreation of meanings:

> The sense of the divine other is cultivated by participation in a coherent ritual system. . . . The ritual system is embedded in, and helps to continually create, a behavioral environment in which participants embody a coherent set of dispositions or habitus. These are the elements that constitute the webs of significance – or of embodied existence – within which the sacred self comes into being. To be healed is to inhabit the Charismatic world as a sacred self. (Csordas 1994: 24)

Believing that the presence of the ship has the potential to make a difference to the outreach communities, that it is a vehicle and a symbol for hope, plays a fundamental role in image self-construction for those who volunteer on board. It bonds and provides a transcendental experience for those to cope with the images of suffering, and their often unattained medical – physical – successes. The point is to keep on, push through with the physical enactment of faith to create healing. A lapse in this would break the intention. 'Being here is a tremendous challenge,' one nurse told me, 'but I've wanted to do this for years. I cannot imagine a more fulfilling way for me to serve God than as I am here right now amongst others who are doing the same.'

Summoning Support

Suffering from a chronic illness can place people in opposition to the 'healthy world', one that they do not feel they inhabit anymore or have access to (Stoller 2004). Depending on the type of illness they have, the new and then enduring experience of being sick can separate them from both their social world as well as their own internal world: unfamiliar bodily functions, appearances, practices and the ever-presence of pain create a dissonance with the lives that they lived before. As Arthur Kleinman and his colleagues state, the individual 'often experiences pain as an intrusive agent: an unwelcome force producing great physical distress as well as moral and spiritual dilemmas' (1997: 5f.).

In much of the literature on pain and suffering there is an emphasis placed on pain resisting language (Good 2007), on it 'devastating the spirit' of those it grips (Cavell 1997) to the point of needing to find other modes of expression. The experience of pain is private and its articulation is public – incompatible with the outside world (Scarry 1985). Veena Das (2001) looks for a language for pain – not just in academic terms but also for those experiencing the suffering (though pain does not equate or necessarily imply suffering). Language and categories may serve to reshape and communicate pain, but they cannot remove or evade it. In which way, then, do individuals reshape their experience in order to cope with their illnesses? How does one transcend the limitations within which one lives and manage to create a support system geared towards the new life circumstances?

One of the nebulous and elusive areas of discussion with ship patients was the subject of the physical pain brought on by their illnesses. While crewmembers, and others in the social circle of my informants, made many assumptions about the pain they were experiencing, direct questions on this subject did not elicit much imagery from the majority of my informants. Instead I was told, 'yes, there was a lot of pain', or 'I took such and such pills against the pain'. Even with Alexandre, as his cancer advanced, what he expressed were the things he was not able to do anymore as a result of his illness, instead of the pain it caused him. Upon my return after spending some time out of the country, I invited him to come visit me in Cotonou before I was able to travel to his village. There was silence in response from him until we organized a trip for me to visit him. When I met him at his home, and saw the decline in his health since the last time I had seen him, I understood immediately. 'This is why I could not come,' he said, indicating his emaciated frame and his face and neck, whose further decay could no longer be hidden by the handkerchief he tied around his head.

If pain, through both physical suffering and mental isolation, removes people from the world, can they be brought back? Perhaps through ritual and language, and a faith-based language at that, that in the process of bringing them into the world also brings others in their world along with them, creating a transformation among all actors (who choose to participate). Those that employ the language and the rituals invest themselves in the experience of the ill person and create a forum for a common familiarity. Being incapable of imagining someone's pain is of a different order than being unable or unwilling to acknowledge it. Participation in these rites represents a mechanism for making suffering visible and legitimate.

Spirituality in any number of forms becomes a mode with which people make sense of their surroundings and those of people close to them. As Csordas (2004) states, alterity is the kernel from which religion is elaborated. In this context, religious faith serves as a method of meaning-making, control and relief, not only for the ill individual, but, by focusing on the illness, also for the members of his or her social network. This external force is kneaded and melded into a catalyst for hope and unity, through the active expression of a belief in God. Language here is not enough; belief and its enactment, using the tools of ritual, language and practice are turned to in order to reconstruct the world caused by illness, as much as for an ill person as for a healthy person needing to make sense of things they are unable to explain.

Away from the ship, the resources involved in creating a sense of healing for both the ill person and his or her network have fluid boundaries. A congregation materializes not only through the physical shape of a church and its inhabitants, or through medical technology as a tool, but belonging comes about also through a conscious choice to manifest one's hope and belief towards a targeted source. Rituals can offer a culturally grounded technique to remove, or negotiate, uncertainty from an experience of sickness or suffering that affect a community beyond the sick person.

The following section examines one of the outlets that Alexandre invited into his world that, by embracing him in return, enabled him to live differently in the face of illness. The ship, for a time, was one such source of support, with its crew invested in his well-being partly for the sake of their own sense of purpose and well-being. On land, he wove together support from many corners, retaining help from people who, in turn, wove him into their lives. As mentioned earlier, just as healthy groups execute social control by stigmatizing 'the other' (sick individuals), 'the other' can try to control and reclaim their social surroundings as well and form their identity through their tailored experiences and the congregation they assemble to accompany them through their journey. The burden of the body itself can be less taxing than the burden of someone else's attitude towards it. In this scenario, those who are ill have summoned a congregation into their lives to redefine themselves as they want to be seen and be able to live in the manner useful to them.

Wednesday Prayers

After the ship left Benin for their next outreach destination in Ghana, Alexandre spent a period of time resting at his aunt's house in Cotonou living with his cousins and his brother. I had the sense that in addition to wanting to be away from the usual place he lived, a small town without much privacy and no escape from the public's continuous tracking of his health status, he wanted to be in a place where he could choose the people he spent time with and the people who supported him. He was fortunate in that he had another place to go live, and was able to exercise more control over the people playing a role – spectator or otherwise – in his life. He spent his days largely at home, impaired by a lack of cash to move freely around the city, but he also visited other relatives and, a couple times a week, a cybercafé, where he would write to nurses he had befriended during his stay on the ship. Twice a week he went to two churches: once on Sundays to the family church, and on Wednesday mornings to another evangelical church for a five-hour service.

These Wednesday services consisted of prayer and singing, interspersed by members of the group giving testimony to how they had been blessed by God in recent weeks. This usually involved a lengthy narrative, punctuated by the drumming and interjections of the lead pastor, with a crescendo to the explanation of how God had worked blessings in this person's life and how grateful he or she was. These services took place not upstairs in the regular church hall, but downstairs in the crypt for greater privacy; unlike with the blatancy of regular church services in the open, passers-by were unlikely to be curious about activities taking place out of view. The idea was for people to feel closer together and to avoid attracting the attention of snoopers curious to listen in on the testimonies of those suffering. The result was an intense heat created by between thirty to sixty people crowded into a small non-ventilated space. Praying raucously increased the muscle of the prayer and helped the believer believe more strongly. I had been told that this was an exclusive service for ill people only, but over time I came to see this wasn't really the case. When she had time, Alexandre's aunt, who had originally introduced him to the group, attended the Wednesday sessions as well, in an act of support for him and as part of a busy schedule of volunteer work that she did.

Fabrice, a thirty-year-old pastor who led some of these sessions, told me that for people who were able to, Wednesdays provided a break within the week to be able to devote themselves to God. Many asked their employers for time off to attend the service, and were granted leave to do so every week. Indeed, some of the participants' testimonies showcased formerly 'stubborn' employers coming around and granting their employees the time off to go to the sessions, given as an example of how God's word was reaching more and more people through the existence of this gathering, and how powerful they were as a group.

This was something I also heard from another pastor who worked with the hospital ship when I brought up the expressiveness of evangelical prayer. He compared it to a supposed 'European' style of Christian worship, and stated that the average Beninois tended not to be concentrated enough to remain focused on their prayer in silence. Shouting, dancing and chanting all forced a person to hear themselves and be aware of their goals, keeping them from slipping into distracted thought and a careless communion with God. Active prayer in groups further heightened the connection, because not only was their prayer more likely to be heard through the power of numbers, but the quality of the prayer would be increased because each individual was able to produce more intensity as a unit.

The communion Alexandre felt with this service, but not the church itself, seemed to be substantial. In his home village he had attended a small church that was quite intimate in nature. He had joined it when he was at his most sick six years before, after receiving a home visit from the pastor who was new to the village. It seemed that he missed the ambiance of the church there and the communion he felt with members of the parish, as he seemed to underline the presence of this new church in his life more than he may actually have attended it. Alexandre was not vigilant about being on time to the service – in fact, he was one of the few people that I observed arriving late and leaving early. Sometimes he sat upstairs in the empty church, perched at the grill that served as a vent down into the crypt, looking in on the congregation. Being involved with the Wednesday sessions, however, held a cachet for him, even if part of the reason was because he knew that I was interested in this area of his activities. Alexandre created a considerable amount of dialogue around group activities that he was a part of, often elaborately recounting the events for weeks afterwards and demonstrating his excitement through decorative speech and high-fives. In an obvious sense, this service fell into a similar category.

All of this conveys one way in which ritual is enacted in order to create a sense of community and hope. What seemed important was not so much the actual outcome of the prayer, but the consistency in carrying it out. Months could go by in which the state of suffering was constant, but even this could ultimately be explained as being part of the process and God's will. Hirobazu Miyazaki's ideas, as he outlined in a chapter of *The Method of Hope* (2004), are useful in elaborating this idea. As he writes: 'the ritual experience of hope and its fulfillment [is] instrumental in the production of hope' (2004: 86). The collective anticipation of fulfilment and the recounting of instances of fulfilment by members of the group, are as important – or even more – than personal fulfilment itself. This is the creation of a unified body with a sick person at the centre, instead of a sick person being outcast, as is so often highlighted in other literature and in other examples of the stigmatization of people with facial disfigurement. It is an enactment of faith that embraces the sick person instead of casting them out, and forms part of the ties that bind people together.

Or, to put it in Pamela Klassen's (2005) words, who describes the experiences of one of her informant's who practiced complementary spiritual techniques (in this case east-Asian practices) alongside Christianity, 'this promised healing was not necessarily curing, but instead a coming to terms with bodily suffering that may or may not eliminate the bodily suffering' (2005: 383). The impact of volunteering to be a part of someone else's healing process rebounds back onto oneself and acts as an aid to 'right' the world, rebalancing the disorder created by illness and misfortune; recasting a lack of comprehension as faith. Hope in this scenario involves believing that one's social investments in one's surrounding worlds will bear fruit, if not immediately, then at some point in the future (Bourdieu 2000).

Turning to these services and opportunities is a way of watching out for oneself and creating a community where there otherwise might not be one. A unifying characteristic among the patients of the ship that I followed was that they actively sought out faith and the ritual enactment of it in order to reconstruct their worlds. The people in their lives that were a part of these networks may have come and gone, but the networks and options remained; the searching for them became an integral part of daily life. In a similar sense, the volunteers on the ship were re-enacting their faith through ritual practices aimed at helping others but also as a means of transforming their own lives as well.

Conclusion

One afternoon ten months after Alexandre's birthday party, and three days after Christmas, Alexandre collapsed and died. He was six weeks shy of his thirty-third birthday, for which party planning had already begun. Serge reached me on the phone that evening to let me know, and through text messages updated me on the burial held the following morning, at dawn, near his parents' compound. Those of us who had become part of Alexandre's life held simple masses in his honour in the following months.

My colleague in Cotonou arranged for a mass held in his name at the hospital church. Even though she had only met him twice, she had followed closely the story of his health. Other ceremonies were held across Cotonou and in his hometown by the pockets of individuals in his life.

One afternoon much later, after some time had passed for both of us to reflect, Serge and I discussed Alexandre's last months, and he filled me in on some of the events I had missed when I was away. According to Serge, old interpersonal difficulties had arisen again in Alex's life. Some months before, he had convinced his aunt in Cotonou to let him expand her beverage-sale business to his village. He threw himself into the vending and managed to build up a client base while living in his usual room at another relative's house. However, the uncle there had something against Alexandre and strategically took over the business, pushing Alex out of the house at the same time. Alexandre became frustrated that something he had built up and worked hard for was being taken away from him, and, when it fell apart, he moved back to his mother's compound in the bush about twenty minutes away from the village by motorcycle. Again, he found himself in an isolated position, geographically and socially alone. Not having money for transportation to head to town, he stopped seeing his friends or going to the doctor for expensive blood transfusions and was unable to vary his daily activities; his health deteriorated.

Serge said he believed that if it had worked out with the beverage sales, Alexandre would still be alive (two years later), even though several years previously doctors and others had doubted he would survive more than a few months with such an acute and grave cancer. According to Serge, being defeated in his venture, where he had been able to socialize and experience success, took away the spirit to ward off the spread of his disease, and thus it had been able to overcome him. Alex's congregation, carefully made up of group and individual relationships he nurtured, who in turn nurtured him, had ultimately failed him when he slipped away out of sight. The weakening corporeality of his support system was not enough to sustain the corporeality of his physical self.

So often in the discussions surrounding healing – whether medical, charismatic, religious and all combinations thereof – talk turns to its efficacy, proof that any results are the product of a definable, unifiable series of steps taken to make a person better. 'Efficacy' can be relevant for the community's health and resuscitation throughout a period of suffering, just as it can be for the impact on an individual life and survival through an illness. Betty's comment in the ward, cited earlier, is relevant here again: 'I want to make someone feel loved,' she said. While the focus previously was on the love expressed and hopefully received, the first part of her statement should not be ignored: 'I want to . . . '. The prayer, energy, emotion and resources she puts forward towards the patients in the ward are not only about them, but also about her being able to work towards her goal of effectuating a personal transformation.

Faith in being able to bring about a change, follow-through in the pursuit of it and the reverberations of these actions all created scenarios of healing, bringing people back into healing when they otherwise might not have been a part of it. Faith continues in the form of those who rally around misfortune, despite, and sometimes because of, the failures encountered along the way.

Isabelle Lange is a research fellow in anthropology at the London School of Hygiene and Tropical Medicine (LSHTM). Her research interests include maternal health (largely questions of quality of care, hospital environments and policy transfer) and identities surrounding healthcare decision-making.

Notes

1. Fortes gives the example of a man finally attaining 'truly complete' personhood only after his father's death, even though he has made steps along the way, such as adulthood rituals and fathering children.
2. Being associated, in a broad sense, with stigma and with people who are stigmatized is often a story of conceptual movement, and self-transformation. Ostensibly, altruistic ideals can be combined with solipsistic or meditational objectives in people who are marked by stigma as well as in those who work with such individuals and groups.
3. http://www.mercyships.org/pages/mercyships-mission (last accessed 20 June 2011).
4. http://www.mercyships.org (last accessed 14 April 2015).
5. I refer to the term 'evangelism' as defined by Unruh: 'sharing the gospel (the Christian message about salvation) by word and deed with people not actively affiliated with Christian faith, with the intention that they will choose to accept and follow Jesus Christ and join a church community for ongoing discipleship' (2005: 34).
6. Transformations are the purpose of the missionary endeavor whether the work is 'religious' or social. At the centre are a proper enlightenment concerning God's will and the import of the sacred events related in scripture: metanoia, conversion, acknowledging Christ and reconciling oneself in Christ through others the realization of Christian faith, hope and love (Burridge 1991: 149).
7. http://africanmercyadventure.blogspot.com/2008_02_01_archive.html.
8. An evangelical tool, *The Jesus Film* is a 1979 film depicting the life of Jesus that has been translated into hundreds of languages.

References

Becker, H. S. 1963. *Outsiders: Studies in Sociology of Deviance.* New York: Free Press.
Berger, J. 2003. 'Religious nongovernmental organizations: An exploratory analysis.' *Voluntas: International Journal of Voluntary and Nonprofit Organizations* 14.(1): 15–39.
Biehl, J., B. Good and A. Kleinman. 2007. 'Introduction: Rethinking Subjectivity.' In Biehl, Good and Kleinman (eds), *Subjectivity: Ethnographic Investigations.* Berkeley and London: University of California Press, pp. 1–33.
Bornstein, E. 2005. *The Spirit of Development: Protestant NGOs, Morality, and Economics in Zimbabwe.* Palo Alto, CA: Stanford University Press.
Bourdieu, P. 2000. *Pascalian Meditations* (translator Richard Nice). Cambridge: Polity Press.
Burridge, K. 1991. *In the Way: A Study of Christian Missionary Endeavours.* Vancouver: University of British Columbia Press.
Cavell, S. 1997. 'Comments on Veena Das's Essay "Language and Body: Transactions in the Construction of Pain".' In A. Kleinman, V. Das and M. Lock (eds), *Social Suffering.* Berkeley: University of California Press, pp. 93–98.
Comaroff, J. 1997. *Of Revelation and Revolution Volume II: The Dialectics of Modernity on a South African Frontier.* Chicago and London: University of Chicago Press.
Csordas, T. J. 1994. 'Introduction: The Body as Representation and Being-in-the-World.' In Csordas (ed.), *Embodiment and Experience: The Existential Ground of Culture and Self.* Cambridge and New York: Cambridge University Press, pp. 1–24.
———. 1994. *The Sacred Self: A Cultural Phenomenology of Charismatic Healing.* Berkeley: University of California Press.
———. 2004. 'Asymptote of the Ineffable: Embodiment, Alterity and the Theory of Religion.' *Current Anthropology* 45(2): 163–85.
Das, V. 1997. 'Language and Body: Transactions in the Construction of Pain.' In A. Kleinman, V. Das and M. Lock (eds), *Social Suffering.* Berkeley: University of California Press, pp. 67–91.
———. 2001. 'Stigma, Contagion, Defect: Issues in the Anthropology of Public Health.' Stigma and Global Health: Developing a Research Agenda; An International Conference. Bethesda, MD, 5–7 September.
Douglas, M. 1966. *Purity and Danger: An Analysis of the Concepts of Pollution and Taboo.* Harmondsworth: Penguin.
Falk, R. 2001. *Religion and Humane Governance.* New York: Palgrave.
Fassin, D. 2012. *Humanitarian Reason: A Moral History of the Present Times.* Berkeley and London: University of California Press.
Frank, A. 1990. 'Bringing Bodies Back In: A Decade Review.' *Theory, Culture, Society* 7: 131–62.
Fortes, M. 1987. *Religion, Morality and the Person: Essays on Tallensi Religion.* Cambridge: Cambridge University Press.

Good, B. J. 2001. *Medicine, Rationality and Experience: An Anthropological Perspective*. Cambridge: Cambridge University Press.
Harding, S. F. 2000. *The Book of Jerry Falwell: Fundamentalist Language and Politics*. Princeton, NJ: Princeton University Press.
Janzen, J. 1982. 'Therapy Management: Concept, Reality, Process.' *Medical Anthropology Quarterly*, New Series 1(1): 68–84.
Klassen, P. E. 2005. 'Ritual Appropriation and Appropriate Ritual: Christian Healing and Adaptations of Asian Religions.' *History and Anthropology* 16(3): 377–91.
Kleinman, A. 1998. *The Illness Narratives: Suffering, Healing and the Human Condition*. New York: Basic Books.
———. 1999. 'Experience and Its Moral Modes: Culture, Human Conditions, and Disorder.' In G. B. Peterson (ed.), *The Tanner Lectures on Human Values*. Salt Lake City: University of Utah Press, pp. 355–420.
Kleinman, A., and J. Kleinman. 1997. 'The Appeal of Experience; The Dismay of Images: Cultural Appropriations of Suffering in Our Times.' In A. Kleinman, V. Das and M. Lock (eds), *Social Suffering*. Berkeley: University of California Press, pp. 1–23.
Kristeva, J. 1982. *The Powers of Horror*. New York: Columbia University Press.
Livingston, J. 2008. 'Disgust, Bodily Aesthetics and the Ethic of Being Human in Botswana.' *Africa: The Journal of the International African Institute* 78(2): 288–307.
Mauss, M. 1985. 'The Category of the Human Mind: The Notion of Person; the Notion of Self.' In M. Carrithers, S. Collins and S. Lukes (eds), *The Category of the Person: Anthropology, Philosophy, History*. Cambridge: Cambridge University Press, pp. 1–25.
Masquelier, A. 2005. 'Dirt, Undress, and Difference: An Introduction.' In A. Masquelier (ed.), *Dirt, Undress, and Difference: Critical Perspectives on the Body's Surface*. Bloomington: Indiana University Press, pp. 1–33.
McCleary, R. M., and R. J. Barro. 2006. *US-Based Private Voluntary Organizations: Religious and Secular PVOs Engaged in International Relief & Development. No. w12238*. Cambridge, MA: National Bureau of Economic Research.
Mercy Ships Philosophy of Missions. 2009. Available at http://www.kopjar.com/christian/update1/mission_philosophy.pdf (accessed 1 April 2009).
Miyazaki, H. 2004. *The Method of Hope: Anthropology, Philosophy, and Fijian Knowledge*. Stanford, CA: Stanford University Press.
Murphy, R. 1990. *The Body Silent*. New York: Norton.
Redfield, P., and E. Bornstein (eds). 2011. *Forces of Compassion: Humanitarianism Between Ethics and Politics*. Santa Fe, NM: School for Advanced Research.
Scarry, E. 1985. *The body in pain: The making and unmaking of the world*. New York: Oxford University Press.
Scheper-Hughes, N., and M. Lock. 1987. 'The Mindful Body: A Prolegomenon to Future Work in Medical Anthropology.' *Medical Anthropology Quarterly*, New Series 1(1): 6–41.
Stoller, P. 2004. *Stranger in the Village of the Sick: A Memoir of Cancer, Sorcery, and Healing*. Boston: Beacon Press.

Thaut, L. C. 2009. 'The Role of Faith in Christian Faith-based Humanitarian Agencies: Constructing the Taxonomy.' *Voluntas: International Journal of Voluntary and Nonprofit Organizations* 20.4: 319–50.
Tomlinson, M., and M. Engelke. 2006. 'Meaning, Anthropology, Christianity.' In Engelke and Tomlinson, *The Limits of Meaning: Case Studies in the Anthropology of Christianity*. New York and Oxford: Berghahn Books, pp.1–37.
Unruh, H. R., and R. J. Sider. 2005. *Saving Souls, Serving Society: Understanding the Faith Factor in Church-Based Social Ministry*. Oxford: Oxford University Press.

Chapter 6

Addiction and the Duality of the Self in a North American Religio-Therapeutic Community

Ellie Reynolds

This chapter explores the relationship between addiction, control and concepts of self in North American society. Specifically, it focuses on a residential community in Portland, Oregon, where residents, many of whom had a history of addiction, came together to share a meditative practice referred to as Ecstatic Contemplation (EC). Ecstatic Contemplation involves a man 'stroking' the upper left-hand quadrant of the woman's clitoris for fifteen minutes. In this chapter, I argue that Ecstatic Contemplation – a practice that teaches submission to an 'external force' – can be understood as analogous to the Alcoholics Anonymous model of alcoholism, which relies on a similar premise of surrender of control. This relationship between self and external forces is examined using Gregory Bateson's theory of schismogenesis. The chapter follows an individual who was central to the community, and examines her relationship to Ecstatic Contemplation and its wider social implications. It argues that while Ecstatic Contemplation may appear to help people manage their addictions, it actually replaces them with an 'addiction' to Ecstatic Contemplation, and reinforces their dependence on an alternative, externalized source of control.

Whilst carrying out fieldwork for a PhD in social anthropology in 2009, I lived with a residential community in Portland, Oregon. The community was made up of roughly fifty people, and was equally divided between the sexes. Some residents referred to the community as an 'intentional community', meaning that members shared certain practices, and had agreed to live together as a 'community' rather than a collection of individuals. The organization (called InTouch, or IT) that the community was based around offered life transformational and coaching classes and courses to members that focused primarily on sexual and emotional relationships. The main practice the residential community was centred around was Ecstatic Contemplation. This was considered a 'meditative practice' rather than a 'sex act', and as such was ritualized. Ecstatic Contemplation is practiced in order to more clearly feel sensation, and particularly sexual sensation. It is practiced in order to become more intimate with the authentic 'energy' that resides solely within the body, and arises as a result of a visceral 'connection' between two people.

The majority of people I lived with in the residence were in their mid-twenties to early thirties and they were predominantly white with varying educational levels and incomes. In one of my first visits to the organization, one of the teachers, perhaps over-exaggerating, told me that she had never met a woman who had been part of the organization who hadn't experienced some form of sexual violence as a child or adolescent. I later found out that many of the women, and some of the men, had previously experienced some form of addiction – to alcohol, drugs or sex – or had suffered from an eating disorder, and most of these people had been involved in a 12-step programme of some form as propagated by Alcoholics Anonymous.

Although there is no articulated relationship between the practice of ECing and previous addictions to substances, it is clear, given the life histories of most of my informants, that in both contexts – ECing and addiction – they enter into a certain, analogous relationship with substance, be it addictive substance or the 'energy' that is accessed through ECing. A dedicated ECing practice, and an addiction to a substance, as well as the healing rituals and rhetoric used in the Alcoholics Anonymous 12-step programme, for example, all involve some kind of reconfiguration of the understanding of personal agency and self-control. As Thomas Csordas (1994) argues, the idea of self-control is a cultural concept fundamental to North American society. What this chapter explores is the complex understandings of self and other that are fundamental to processes of addiction and to processes that are considered healing. The chapter follows the story of Laila, her past as an addict, and her present as a EC practitioner and teacher within the organization. It explores how, if at all, her sense of self as an addict has been transformed by the supposedly healing practice of ECing.

Ecstatic Contemplation

In the words of the founder of InTouch, the precise method of Ecstatic Contemplation is for the woman to

> butterfly her legs open, he [the male stroker] would put the pillow underneath her knee and then the other knee would rest over her leg. See the idea is that she can relax all of her muscles. He would take one hand, put it beneath her behind, and rest his thumb at her entroitus. The entroitus is the hole, the area where, if you have intercourse, there is penetration. And both people would draw their attention to the place where the connection was being made. Just keep drawing the attention back, no matter what distracting thoughts come up, no matter how ecstatic the sensation, whether or not you're feeling anything at all, it doesn't matter. Up down, up down, up down and to really keep your focus there. The only question that I have to attend to in that moment is 'What is the sensation at the point of connection?' (YouTube video on ECing)

The point of connection is the man's 'stroking' finger (his left index finger) and the upper left-hand quadrant of the woman's clitoris (which has the greatest number of nerve endings). The up-down stroke (otherwise known as the 'bread and butter stroke') is the most basic one taught, and is designed to produce the most 'sensation' for both men and women. At present, according to the InTouch website, it takes ten days to 'feel' anything during an EC. When I was living in the community it was generally acknowledged that, with regular practice, one would start 'feeling' within two to four weeks, but would not become proficient before six months of at least daily practice.

ECing connects the practitioner to their 'orgasm', which is considered to be a gateway to a source of power or energy that is omnipresent, and with which all individuals are imbued. For IT and its practitioners, orgasm is in constant motion, particularly during an EC and manifests itself within the body through physical sensation. Sexuality is a way to access this power source and draw from it. There is a clear relationship identified between neurophysiology and accessing the Divine.

Robert Fuller notes the rise of the concept of healing 'energies' in the United States over the past couple of centuries. He argues that 'practices such as acupuncture, tai chi, yoga, Ayurvedic medicine, shiatsu and many massage therapists also understand their therapeutic benefits in terms of the presence and casual power of energies not recognized by Western science' (2004: 383). Fuller further argues that these alternative healing systems are 'prime carriers of the American metaphysical tradition' (ibid.).

During a recent interview, a friend of mine, talking about his transformation through ECing, explained:

> At first I saw it as something logical and therefore neurological. I held ECing as a way to retrain the parts of my brain that had numbed to sensation. It's like the physiological or psychological obstacles I had put in place to keep sensation at bay were eroded over time from ECing.
>
> Over time, as I've come to know the connection I have with my EC partner and how our orgasm becomes shared in the experience, I have come to see it as more of a spiritual practice. There is an energetic body that is created during an EC that is distinct from the two individuals ECing.
>
> As an atheist, I've come to appreciate that there's an energetic force in the world that is beyond me. It feels good to be able to acknowledge something bigger than myself without having to personify or deify it. (Male informant Jesse, personal communication)

A well-developed ECing practice gradually leads people to be in a perpetual state of orgasm, or to be constantly 'turned on' through the development and honing of their limbic system (the part of their neurological system that regulates emotion and sexual desire). The effect of this is felt in a number of ways. Practitioners notice an increase in physical energy, an increase in awareness of sensations in the body and, ultimately, an increase in the awareness of the sensational states of other people.

Healing within a broader religious or spiritual context can offer many opportunities to participants. Winkelman notes:

> Religious psychosocial and behavioral pathways include a larger social network and support system, social ties that can provide a network that enables people to cope better with problems, reducing stress, depression, and self-destructive behaviors and providing needed resources. Religion also has psychodynamic effects through beliefs and their consequences for peacefulness and self-confidence that can help reduce stress, anxiety and conflict, creating emotionally tranquil states. (2004: 456)

Advanced practitioners can access the source of power through their orgasm whenever it is required. My friend described his new ability:

> My renewed ability to feel turned the color back on in my awareness. I think of it as the difference between when you are 'in love' and when you aren't. When you're in love, everything seems so vivid and exciting. When you aren't, the colors are flat and dull, and everything that's wrong in the world stands out to you in 3D. That's how it was for me being celibate all those years. The world was flat, dull and full of problems and morons. As my ability to feel turned back on, I could see the lusciousness of the world around me, I could tap back into the excitement of a beautiful day, even when it was raining. (Jesse, personal communication)

ECing leads to an ability to not only empathize with the imagined emotions and sensations other people are holding within their body, but to actually feel them. This appears, for most of the people I spoke to, to be the basis of spiritual experience. My friend said:

> I can connect deeply to the person next to me, because I can feel their joy, their pain, their desire. Walking around in the world becomes an infinitely more joyful/painful experience. Yes, my capacity to feel despair and pain increases as well, and I wouldn't have it any other way. It's an honor to truly feel other humans, even when that is desperation, grief and loneliness. In fact, it's better when I can feel that, because I can relate to it, and the connection becomes even deeper. (Jesse, personal communication)

Addiction and Ecstatic Contemplation

Laila was born into a middle-class Jewish family in Dallas, Texas, in the early 1980s. At four years old, her grandfather started sexually abusing her. At eight she developed anorexia nervosa, and in her early teens she started using, and became addicted to, alcohol, cocaine and amphetamines. After leaving high school she decided to get sober and moved to Israel, where she became Orthodox and went into recovery. On her return to the United States three years later Laila got a degree in addiction studies from a university in Texas and started work as an addiction counsellor in a rehabilitation centre. Her aim was to provide art psychotherapy to young women struggling with addictions following sexual abuse and violence.

One day Laila learnt that her boyfriend was cheating on her with a stripper. She arranged to meet the woman with whom her boyfriend – Aaron – had been cheating. Laila described how she had been standing in a kitchen, holding a cup of herbal tea when she met the stripper, Liana. Liana had turned to her and said 'There is a place in Portland, Oregon where they teach mindfulness around sexuality. You need to go there.' Laila described watching in slow motion as she dropped the mug of tea on the floor, and her arms broke out into hives. Following this encounter, Laila and Aaron went on a one month road trip and ended up at InTouch in Portland. I later asked Laila what Liana had seen in her to make her tell her about InTouch. Laila replied that she had seen her 'power'.

Laila had an innate power to 'feel' people, that is, to sense their inner emotional states and desires – essentially she could tap into their limbic systems. This was a power she was born with and, as such, considered it a gift from God. At the time I was living at the InTouch residence, Laila, by now a senior teacher, was going through a relapse of her anorexia and had started seeing a therapist and nutritionist weekly. She would often forget to eat, claiming she did not feel hungry, or would work through meal times to distract herself and others from her lack of desire to eat. Her eating disorder affected her relationship with others. I was sitting with her one day in the centre and one of the residents approached us. He very slowly and gently asked Laila if she had eaten lunch that day. She thanked him and replied that she had. I felt offended by the question and asked her whether she ever got angry that people felt they had the right to ask about her eating habits. She said that she did but lost the right to privacy about her body and food when she became anorexic.

The way Laila spoke about herself in relation to her eating disorder was as if she was split in two; sometimes she was guiding her thoughts, feelings and decisions and sometimes her eating disorder was. When she spoke about it I often got the sense that the disorder was female, and wild, malevolent and vicious. In an email she sent to me after I left the field, Laila wrote in response to my inquiry about her therapy session: 'Yesterday was like a heavyweight champion boxing match. Man my eating disorder is mean.' It also affected the way she experienced her body. Whilst describing a feeling of relief to me when we were out walking she said that she felt like a weight had been lifted from her stomach.

Laila was a proficient ECer. She had a twice-daily ECing practice and had been developing this for the two years she lived at InTouch. Laila told me that she had 'freaked out' when she had first ECed, and that she had felt no sensation. Eventually she started feeling a small amount of sensation in her clitoris, the sensation then spread slowly to the rest of her body after ECing for several months. ECing was important to her and she was a well-respected practitioner. She taught the practice course and gave live EC demos.

An Addiction to ECing?

Whilst sitting in an Indian restaurant in New York, Laila told me about her relationship with ECing. Her power came in the form of energy. This energy gave her the ability to function, as well as 'feel' others. It gave her the motivation she needed to fulfil her purpose, which was to bring awareness to women's sexuality, and guide them through their exploration of it. Her energy allowed her to champion InTouch and make it central to her life. Previously, I had asked how ECing had affected the energy she had in her body. She explained that ECing had grounded the energy, making it feel more solid in her body and more accessible. It also allowed her to rein it in and have more control over it than she would otherwise. Over the meal, I asked her what would happen if she stopped ECing. She told me her energy would eventually run out, leaving her unwell, despondent and deflated. She estimated that if she discontinued her ECing practice she had enough reserves of energy in her body to keep her going for several years before they started to become so depleted that they would impact her body.

Meeting Laila in New York, a year after I had left the field, I was shocked by how central InTouch was to her life, and how little she had outside it. For her, life revolved around promoting InTouch, supporting clients through their practice, activities such as yoga and meditation (which were other body processes which allowed her to manage energy), the importance of food (monitoring her eating disorder and maintaining a gluten-free diet) and managing a complex relationship with her family. All her conversations and activities were centred around InTouch and she was working or on call twenty-four hours a day. At times, whilst walking around the city, I had to lead her around by the arm while she responded to phone calls, emails, and text messages from colleagues or clients. She had little to no contact with people who weren't involved with InTouch and I noticed that she often cut ties with people who were no longer considered beneficial to InTouch, despite enjoying a close relationship with them when they had been involved.

I spoke to a mutual friend in London recently who had attended a coaching course at InTouch in Portland a couple of weeks before. He described to me the recent concern there around Laila's quite ruthless behaviour towards potential clients. She had been accused of bullying and coercing people into taking courses and giving money they did not want to give, and it had been noted that people found her fickle and false in her relationships, only maintaining those that were beneficial to the organization. When questioned about this she would respond that her loyalty lay with the organization rather than individuals, and that she would do anything to see it succeed.

Addiction

Psychologist Stanton Peele (1985) argues that people can become addicted to any experience that modifies mood, meaning that anorexia, bulimia and overeating fall well within this category. In *Epidemics of the Will* Eve Kosofsky Sedgwick (1992) has outlined the development of 'addiction' since its 'invention' in the late nineteenth century. Since this time, she argues, more and more substances and actions have been branded 'addictive', and she traces the movement of this characterization from opium, to alcohol, to food, to exercise and, latterly, to work (ibid.: 582–84).

The idea of addiction to a foreign substance, even one as benign as food, in the form of anorexia, bulimia or overeating, deconstructs the very notion of 'addiction', Sedgwick argues:

> If addiction can include ingestion *or* refusal *or* controlled, intermittent ingestion of a given substance, and if the concept of 'substance' has become too elastic to draw a boundary between the exoticism of the 'foreign substance' and the domesticity of, say, 'food', then the locus of addictiveness cannot be the substance itself and can scarcely even be the body itself, but must be some overarching abstraction that governs the narrative relations between them. (Ibid.: 583)

The abstraction that governs the space Sedgwick refers to as 'healthy free will', or rather, its lack, whereby substances (food, alcohol) are understood to be imbued with mystical properties such as 'consolation, repose, beauty, or energy' and when imbibed, operate on the self 'here self construed as lack' (i.e. the self is considered to be lacking in some way that only the consumption of the magical substance can reiteratively fulfil). This view – of lack or loss that becomes spuriously fulfilled through the consumption of destructive substances – is one present in contemporary lay and medical discourses on addiction (ibid.).

Rejecting this explanation, Sedgwick argues that now, any substance, any behaviour – even any affect – may be pathologized as addictive. 'Addiction, under this definition, resides only in the *structure* of a will that is always somehow insufficiently free, a choice whose volition is insufficiently pure' (ibid.: 584). In other words, those activities that appear to be products of the choice and free will afforded to us within the individualist climate of late capitalism – shopping, sex, eating, work, exercise – are precisely the ones that are being pathologized as addictive. Along with the hypostatization of 'free will', there was, inevitably, the hypostatization of compulsion (ibid.). Compulsion is always lurking behind volition, threatening to take over. Sedgwick invokes Nietzsche's work as she writes that '*all* tied to the bizarrely moralized imperative for the invention of a Will whose value and potency seem to become more absolute as every grounding possibility for its coming into existence breathtakingly recedes and recedes' (ibid.: 586).

Just like the hypostatizing of free will with the Reformation, Sedgwick argues for an historical determinant for 'addiction'. She identifies the location of this determinant in 'the peculiarly resonant relations that seem to obtain between the problematics of addiction and those of the consumer phase of international capitalism' (ibid.: 587). From the nineteenth century until the present there has been a sudden expansion in international trade, and increasing access to exotic, luxury goods, including such 'addictive' goods as heroin, opium, alcohol and tobacco.

Excessive consumption of such goods became pathologized when a moral distinction was made between 'natural' (e.g. food) and 'unnatural' (e.g. drugs) desires. 'Natural' desires were 'needs', and those unnatural were 'addictions'. Pathologization was also applied to the scale of consumption; Sedgwick here quotes Oscar Wilde's *The Picture of Dorian Gray*: 'Anything becomes a pleasure if one does it too often.' The puritanical fear of excess turns consumption into addiction, and volition into compulsion.

Sedgwick ends by juxtaposing the American notion of 'freedom' with the lack of actual freedom, and the authoritarian invasion of the state into everyday lives, and suggests that contemporary 'addictions' are ways to resist the imposition of moral authority into desires for consumptive ingestion, even excessive consumption, and the loss of agency through pathologization paradoxically restores it.

Alcoholism

In the rhetoric around 'alcoholism', and particularly that used by Alcoholics Anonymous, alcohol is a substance that can damage the system or network of social relations an individual is embedded in (Swora 2001). In the AA view, alcoholism is an incurable and progressive disease characterized by loss of control over drinking. It is also characterized as being ground into the bodily substance, as it is an incurable affliction. As Moira Swora argues, 'to be an alcoholic, as AA sees it, is to carry a certain identity grounded into one's essential bodily being' (ibid.: 5). Recent arguments that there is a genetic component to alcoholism suggest it is the result of a shared bodily substance (DNA). Swora points out that 'the disease concept of alcoholism acts to bind alcoholics who join the fellowship into a moral community of who are in a sense kin because of a sense of consubstantiality. They see themselves as sharing both a legacy of suffering and body substance and thus social identity' (ibid.).

So, while the identity of an alcoholic is at once the rejection of a substance, it is also an acknowledgment of the power it still has over the self (and will continue to have, thanks to the 'incurable disease' metaphor), which is reiterated endlessly. Alcoholics and recovering alcoholics who subscribe to an extent to the AA rhetoric also construct their new identities through different transferable and non-transferable substances, namely 'disease' and 'DNA'.

Paul Antze has described the social form that has resulted from a specific discourse and relationship to substance:

> AA as a social form has a certain 'totemic' quality. The group's name, its central preoccupations, its solidarity and power to confer identity all spring from a single substance which members collectively abjure. Furthermore, as with most totemic societies, members avoid the substance not out of moral qualms, but because they see themselves as belonging to a category of persons uniquely endangered by it. (1987: 150)

For some, alcohol as substance has its own agency; it becomes the enemy and it requires constant vigilance to overcome its advances.

Theorists examining the sociological aspects of alcoholism, such as the anthropologist Gregory Bateson (1985), have argued that alcohol, as substance, is a symptom, or representation, of the individual's inability to relate properly to the social world. For Bateson, the alcoholic perceives himself as being separate from, and in opposition to, the world, rather than being situated within it, and specifically within a wider network of social actors. AA works by providing a wider network, and alternative substances (concepts of disease, God, DNA) through which alcoholics can configure social relationships and idioms of belonging in 'proper' ways.

AA teaches alcoholics what constitutes proper and improper relationships within the AA and within the broader context of US society (Swora 2001: 7). The AA view of the person is one who accepts his limitations, who relinquishes control to a Higher Power (God) and who has the capacity to be honest with himself (and with God), rather than to be in denial (ibid.: 8–9). In *The Sacred Self* Thomas Csordas (1994) identifies control as a major cultural theme in North American society. The hallmark of addiction is loss of control over a substance.

In *Steps to an Ecology of Mind* Bateson (1985) uses the idea of alcohol addiction and its subsequent treatment through AA to further his theory of schismogenesis. Schismogenesis implies 'creation of division', and Bateson uses it to describe a development in certain types of social relationships or social behaviour between individuals or groups. The first – complementary – describes a relationship between two categorical unequals, such as between God and a worshipper. The second – symmetrical – describes a relationship between two rivals, such as the arms race between the United States and the Soviet Union. Bateson argues that both forms of social relationship are liable to progressive changes or escalation, which he refers to as schismogenesis. The two relationships can coexist and impact on each other in various ways.

Bateson, like Sedgwick, notes that in the lay and professional understandings of alcohol addiction there is an assumption that there is some pathology or 'lack' in the alcoholic's sober life that pushes him to alcoholic intoxication. Bateson writes, 'if his style of sobriety drives him to drink, then that style must contain error or pathology; and intoxication must provide some – at least subjective – correction of this error' (ibid.: 310)

'Contemporary' understandings of alcoholism also assume that the alcoholic is addicted because of a weakness of character – that is, he should be able to resist alcohol and be the 'captain of his soul' (ibid.: 312) but is too weak to resist temptation. This, Bateson argues, is a result of the pervasive Cartesian dualism of mind and body in Western society, or, 'rather, between conscious will, or "self", and the remainder of the personality' (ibid.: 313). The genius of AA, according to Bateson, was to break up the structuring of this dualism (that 'mind' or 'conscious will' is transcendent, rather than immanent, in relation to 'matter' or the unconscious) with the first two of the twelve steps:

1. We admitted we were powerless over alcohol – that our lives had become unmanageable.
2. Came to believe that a Power greater than ourselves could restore us to sanity.

Bateson argues, against Descartes, that 'the mental characteristics are inherent or immanent in the ensemble as a *whole*' (ibid.: 315). The system of the person is instead internally interactive. Mind is immanent in the whole system – brain plus body, and mind is immanent in the larger system – man plus environment. The 'Occidental' (Euro-American) conception of the self has resulted in the idea that all agency is situated within the self, specifically within the mind, rather than being dispersed within a system. For example, a man cutting down a tree with an axe will modify each stroke of the axe according to the cut face of the tree left by the previous stroke. This self-corrective process is the result of a total system tree-eyes-brain-muscles-axe-stroke-tree, 'and it is this total system that has the characteristics of immanent mind' (ibid.: 317). The Euro-American self would see this as an act (muscles, axe, stroke) of agency (self) on an object (tree), and would therefore say, '*I* cut down the tree.'

Alcoholics should also be understood as being part of a wider system of networks. Bateson points out that consciously peripheral mental phenomena (actually abstract principles that guide and govern human behaviour) are attributed to the body as 'feelings' in Euro-American understandings of the self. This is compounded by the fact that they are often experienced viscerally as physical sensation.

For Bateson, a central characteristic of alcoholics is 'pride', taken to mean 'I can. . . .' (as in, 'I can stop drinking' rather than 'I succeeded').

Success destroys the challenge, so someone who *can* stop drinking no longer feels the need to stop drinking. Pride in this context is linked with 'risk-taking', which Bateson (1987) describes as: 'I can do something where success is improbable and failure would be disastrous.' Pride in this formulation assumes another, an observer, and the alcoholic's relationship to the observer can be characterized as either complementary or symmetrical. Bateson sees symmetrical (competitive) relationships in normal Western drinking habits, as a result of friendly masculine competition. As the alcoholic starts becoming addicted to alcohol, his drinking starts being viewed as a weakness by those around him and his relationship to alcohol becomes increasingly complementary. He attempts to prove he is dominant over alcohol and that he is 'captain of his soul'. Here he falls into the Cartesian dualism fallacy outlined above, that the mind is transcendent of the body.

When the alcoholic battling the urge to drink *does* succumb, he is transformed:

> His anxieties and resentments and panic vanish as if by magic. His self-control is lessened, but his need to compare himself with others is reduced even further. He feels the physiological warmth of alcohol in his veins, and, in many cases, a corresponding psychological warmth toward others. He may be either maudlin or angry, but he has at least become again a part of the human scene. . . . In ritual, partaking of wine has always stood for the social aggregation of persons united in religious 'communion' or secular Gemütlichkeit. In a very literal sense, alcohol supposedly makes the individual see himself as and act as a part of the group. That is, it enables complementarity in the relationships that surround him. (Ibid.: 329)

At this stage, the alcoholic's relationship to alcohol has become complementary rather than symmetrical. He is not battling it, but succumbing to it. He is submissive to alcohol's dominance.

The next stage whereby an alcoholic's relationship to alcohol becomes complementary is when he joins Alcoholics Anonymous and takes on the first two steps, outlined above. The alcoholic enters into a complementary relationship with both alcohol, and with a Higher Power, and is transformed in the dualistic relationship with his own physical desire for alcohol. He recognizes himself as part of a wider system, and in relationship to God, as he understands Him to be.

This holistic transformation is a key characteristic of healing in religious or spiritual traditions in which death and rebirth imageries are frequent. They can perhaps be compared to Mircea Eliade's description of initiation rites, which involve a transformation of worldview. Such rites communicate a sense of taking part in a new and more genuine reality, creating an individual who considers themselves 'a being open to the life of the spirit' (1965: 3).

Control and Surrender

As mentioned earlier, Thomas Csordas (1994) has argued that 'control' is a major psychocultural theme in North America. This theme is also present in the discourse of demonic possession among the Christian Charismatics he writes about in *The Sacred Self*. Csordas understands demons that possess those who attend rituals to be healed as an externalization and objectification of the negative aspects of the self. Csordas argues that for adults to be considered healthy in North America they must be thought of as having control over their behaviour and feelings. Within the healing system of the Charismatics, demonic presence is discerned through a loss of control over thoughts and feelings. This is externalized and objectified by the patient as 'something inside me' and is reminiscent both of Laila's female eating disorder, lodged inside her mind and body, and Bateson's (1985) understanding of the subconscious, exerting its influence on the conscious mind ('self') and preventing it from becoming the 'captain of the soul'.

Csordas (1994) notes the metaphors used in the cross-cultural anthropology of demonic possession and argues that while the metaphor of interiority/exteriority (as in 'I am possessed because the demon or spirit is inside me') is used in order to gain true understanding of this cultural phenomenon, it must be understood according to the phenomenological metaphor of control and freedom. Demonic possession is frequently characterized as a demon (or the negative aspect of the self) hanging onto or latching onto the person at the site of an emotional wound and are phenomenologically experienced as such.

Regarding the ritual casting out of the demon and reconstitution of the self, Csordas argues against the separation of the 'thing expressed' and its 'expression':

> I would suggest that the 'thing expressed' that 'does not exist apart from the expression' is in this case not the cultural object, the evil spirit. What is expressed is a threshold of intensity, generalization, duration, or frequency of distress that is transgressed – there is too much of a particular thought, behavior, or emotion. The phenomenology of the process defined by discernment, casting out, and manifestation can be summarized by the formula, 'I have no control over this. It has control over me. I am being released.' (Ibid.: 220)

That is, the demon being cast out is merely a metaphor for the transcendence of emotional distress (the negative aspect of the self) and the relinquishing of control. The self is no longer battling against its fragmented negative, but the two and their fragmentation are transcended and reconstituted within ritual, therefore, just like within Alcoholics Anonymous and the relinquishing of control to a higher power, the theme of control loses its cultural importance and visceral power.

Both Swora (2001) and Bateson (1985) note the importance placed by AA on honesty in the process of healing. As Swora writes, 'the proper relationship to God is dependent upon the capacity to be honest with oneself, and idea related to the notion of "denial"' (2001: 9). These notions of secrecy, denial and honesty are tied in with ideas of control and freedom – that only when the alcoholic acknowledges their lack of control can they be truly honest with themselves and others. Failure of the AA programme is due to a lack of ability to be honest with oneself. Again, there is the idea here that the self is fragmented, and that one part of the self can lie to the other about the ability to control desires and cravings, leading to the idea of a constant battle between fragmented parts of the self. Secrecy and dishonesty occur once there is a loss of control over a substance, and a subsequent attempt to gain control. The addict is dishonest not only to himself but to those around him about his consumption (or not) of substance. One of those relationships is with God, or the Higher Power.

The premium placed within North American culture on self-sufficiency and self-control, as well as notions of personal and consumptive freedom that have emerged since the Reformation and in late Capitalism, have led to a dominant cultural conception of the ideal modern self as impermeable to external forces, and subject only to its own moderate, and controllable, desires. The disjuncture between these states (of control and freedom) have led to the development of addictions as individuals lose a sense of their own control in their ability to consume and instead surrender to an externalized agency (such as food or alcohol) that now controls them. This externalization of agency is a result of an unwillingness or an inability to manage the delicate balance between self-control and freedom, and so the locus of control is shifted. In anorexia, the locus of control is also shifted to the externalized (but embodied) 'eating disorder', which the individual ultimately surrenders to. What starts out as an exercise in extreme control becomes a surrender to starvation.

It is this rather ambiguous and paradoxical externalization of agency and control, to the extent of personification, within addiction that interests me here. And it is this precise same externalization that takes place during the healing rituals of Alcoholics Anonymous and the Charismatics, as described by Csordas (1994), whereby power over the self is externalized and located in the Higher Power.

Addiction and Dualism

Considering the above, whereby addiction is regarded as the loss of will, or loss of control over the consumption, or not, of a substance, it is clear that Laila no longer has a choice over her ECing. Without it she would become ill, lose her energy, revert back to other addictions, go insane, and lose her purpose.

However, rather than understanding this as a dependence or addiction affecting an individual, in order to make sense of it it must be understood within the context of the social form and structure of InTouch. There is a dualism that is fundamental to the organization of InTouch: in order to become powerful within the organization, one must submit to the energy – and the form it takes – produced by ECing. Given that the energy that dictates social form within the organization is directly experienced as visceral force, it might be useful to turn to Durkheim at this point.

In the *Elementary Forms of Religious Life* Durkheim (1995) famously argued that religious concepts and experiences, and particularly what he refers to as 'force', are grounded in the reality of the social groups that represent them. The experience of 'force' as a religious concept – for example, the experience of guilt by Roman Catholics – is a symbol of social power exerted over individuals by the collectivity. In his exploration of animistic and totemic religions, Durkheim points to an 'anonymous and impersonal force' that is said to be found in each member of the community, but is not to be confounded with any of them. It exists independently of any individual and precedes and survives them. Durkheim says of Melanesian *mana*, 'It is a power or influence, not physical and in a way supernatural; but it shows itself in physical force, or in any kind of power or excellence which a man possesses.' Durkheim compares the relationship of society to the individual with the relationship of the divine to the worshipper. In both contexts, the individual is expected to behave in a certain way according to certain divine or civic principles.

This divine or civic authority repels representations that contradict it, and keeps them at a distance, it 'commands those acts which realize it' through the emission of psychic energy that it contains.

> The collective force is not entirely outside of us; it does not act upon us wholly from without; but rather, since society cannot exist except in and through individual consciousness, the force must also penetrate us and organize itself within us; it thus becomes an integral part of our being and by that very fact this is elevated and magnified. (Ibid.: 227)

The concept of energy experienced and felt within InTouch appears almost identical to the Melanesian *mana* that Durkheim writes about in that it is a viscerally felt energy and force that reproduces the ideology and so the hierarchical structure of InTouch. A result of this is that the power that people hold is a conditional power and dependent on the reproduction of the ideology. In a sense, these people have a lot less agency than people who are less powerful because in order to maintain their position, they must remain bound by the structure of the ideology.

Because, as Durkheim argues, social form and structure within religious context is experienced as 'force', and as physically felt moral imperative, the body and physical wellness are implied and impacted. Given that many women, including Laila, who start ECing and rise within the hierarchy of power at InTouch come from marginal, abusive backgrounds, InTouch can be seen as a clear opportunity to gain some social power, respect and self-worth. The emphasis on women's bodies and sexuality within the organization means that this becomes embodied and so respect, self-worth and power over others (according to the InTouch hierarchy) are experienced viscerally. Giving this position up means that women will lose their place in the hierarchy, and this will have an impact on their body, as illustrated by Laila when she spoke about the physical impact of not ECing.

ECing within the context of InTouch was experienced as healing and as a way to overcome addiction because individuals understood themselves as whole – that is, they had complementary relationships (*qua* Bateson) with themselves, rather than an external substance or organization. Practitioners were taught that the energy they had came from within their bodies and was natural. As such, it was superior, 'cleaner' and more genuine than their minds, which had been shaped by a warped society. To heal they had to be 'true' to themselves and their bodies.

In fact, practitioners had a complementary relationship to InTouch and its hierarchical structure. If addiction is understood as a viscerally felt, physical dependence on an external agent, rather than the reliance on the self, then it is quite clear that practitioners become addicted to ECing and the social power it symbolizes and to which is allows access.

Ellie Reynolds is a research associate at the Florence Nightingale Faculty of Nursing and Midwifery at King's College London. She undertook an MRes and PhD in social anthropology at University College London. Her main research interests lie in the relationship between embodiment and social structure, with a particular focus on substances and boundaries. She is currently part of a research team carrying out a national evaluation of Schwartz Rounds, an intervention to enhance well-being and compassion in healthcare staff in the UK health service.

References

Antze, P. 1987. 'Symbolic Action in Alcoholics Anonymous.' In M. Douglas (ed.), *Constructive Drinking: Perspectives on Drink from Anthropology*. New York: Cambridge University Press, pp. 149–81.
Bateson, G. 1987. *Steps to an Ecology of Mind: Collected Essays in Anthropology, Psychiatry, Evolution, and Epistemology*. Northvale, NJ, and London: Aronson.
Csordas, T. J. 1994. *The Sacred Self: A Cultural Phenomenology of Charismatic Healing*. Berkeley and London: University of California Press.
Durkheim, E. 1995. *The Elementary Forms of Religious Life* (translated and with an introduction by K. E. Fields). New York and London: Free Press.
Eliade, M. 1965. *Rites and Symbols of Initiation*. New York: Harper and Row.
Fuller, R. 2004. 'Subtle Energies and the American Metaphysical Tradition.' In L. L. Barnes and S. S. Sered (eds), *Religion and Healing in America*. Oxford and New York: Oxford University Press, pp. 375–86.
Sedgwick, E. K. 1992. 'Epidemics of the Will.' In J. Crary and S. Kwinter (eds), *Incorporations*. New York: Zone Books, pp. 582–95.
Swora, M. G. 2001. 'Commemoration and the Healing of Memories in Alcoholics Anonymous.' *Ethos* 29: 58–77.
Peele, S. 1985. *The Meaning of Addiction: Compulsive Experience and its Interpretation*. Lexington, MA, and England: Lexington Books.
Winkelman, M. 2004. 'Spirituality and the Healing of Addictions: a Shamanic Drumming Approach.' In L. L. Barnes and S. S. Sered (eds), *Religion and Healing in America*. Oxford and New York: Oxford University Press, pp. 455–70.

Chapter 7

Religious Conversion and Madness

Contested Territory in the Peruvian Andes

David M. R. Orr

Like much of Latin America, in recent decades the Andean countries have seen shifts in the religious affiliations of a significant minority of their populace. While exact figures are contentious, Timothy Steigenga and Edward Cleary (2007: 12) estimate that around 10–15 per cent of the population of Peru is now Protestant. This represents an important inroad into the overwhelming predominance previously enjoyed by Catholicism. Various – mainly evangelical – Protestant churches are now a fixture of Andean towns and highland villages. Their pastors and congregants have become important actors in a variety of social arenas, from education to community development (de la Piedra 2010; Scarritt 2013; Olson 2006). One such arena that has, however, been under-studied by Peruvianists is health.[1]

The offer of divine healing is characteristic of Pentecostal churches worldwide, and has been a feature of Peruvian evangelicalism since it first achieved prominence (Kamsteeg 1991, 1993). Though these churches are far from attaining the authority in matters of healing that they enjoy in, for example, Ghana (Read, this volume), I soon found during my fieldwork that the prospect of evangelical input was commonly suggested, discussed and contested in relation to many illnesses. This was particularly evident with mental illness, the main focus of my study. Many sufferers, and members of their families, wondered what might be gained by resorting to the *hermanos* (Sp.: 'brothers'), as evangelicals are commonly known, and what the implications would be. Illness and madness raised

Notes for this chapter begin on page 150.

questions around religious loyalties and conversion; in turn, religious loyalties elicited judgements on madness and its causes. In this chapter I use mental disorder as a lens to explore how the Andean peasants and health professionals with whom I worked positioned themselves in relation to the Catholic and evangelical churches, and to the possibility of transition from one to the other. The ethnographic account indicates that greater engagement with these religious institutions would considerably strengthen mental health work in the region.

Conversion

Conversion from Catholicism to evangelical Protestantism in Latin America has been much studied.[2] Peruvian conversion patterns have reflected both wider trends across Latin America and the country's own distinctive recent history. The Andeanist literature on the topic has emphasized how evangelicalism provides a possible response to sweeping societal changes that have occurred throughout the region. Some have suggested that it is more adaptable to the modern market economy as it facilitates a more individualist outlook that permits greater opportunities for capital accumulation. Notably it allows converts to extract themselves from the web of collective contributions and obligations of the fiesta economy associated with Catholicism, though often now in decline among Catholics themselves (Canessa 2000; Gamarra 2000). Others have pointed to how it provides a basis for co-operation between households in the absence of earlier social institutions such as the fiesta (Seligmann 1995: 152; Scarritt 2013). Rural–urban migration played a significant role in conversions, as evangelical churches often offered unparalleled welcoming and supportive networks for peasants newly arrived in the cities, some of whom later returned to their communities to spread the gospel (Muratorio 1980; Paerregaard 1997). Conversion also provided an avenue by which to redefine indigenous or peasant identities by disavowing drunkenness, ignorance and machismo; identifying with these aspects of evangelicalism could both raise one's status in the eyes of the dominant national society (Muratorio 1980; Seligmann 1995: 152; Canessa 2000) and provide a 'project of the self' (Lazar 2008: 160) to aspire to and around which to reformulate traditional gender expectations and family roles. Encouragement to read the Bible for oneself and the ecstatic experience of a direct relationship with God have appealed to many, especially those who experienced the Catholic church as distant, or even oppressive (Seligmann 1995: 153; Lazar 2008; Theidon 2014). Scholars have also suggested that the narratives of morality and salvation that characterize evangelicalism resonated particularly with communities who bore the brunt of Peru's violent armed conflict during the 1980s and 90s and who found in direct engagement with Biblical narratives a way of making sense of their own experiences of suffering, the need for atonement and a society that seemed in need of a new moral compass after losing itself in savage bloodshed (Gamarra 2000; Theidon 2014). This brief summary, while not exhaustive, conveys some of the diversity behind conversion trends in the Andean nations and suggests something of how instrumentality and meaning may both play an intertwined part in individual decisions.

A slightly different, though entirely compatible, perspective comes from a body of research into Latin American conversion that has emphasized changes in religious affiliation as a response to the pressures of ill health, deprivation and poverty. John Burdick, working in Brazil, drew on the notion of Pentecostalism as a 'cult of affliction' (1993: 67) to develop a model of religious choice that derived directly from medical anthropology and analyses of medical pluralism (ibid.: 8). This was taken up by Andrew Chesnut, who suggested that 'the dialectic between poverty-related illness and faith healing' was central to the spread of Pentecostalism (1997: 6). Arguing that a life crisis stemming from illness is very often the variable that determines why one Brazilian struggling in poverty converts and another does not, he makes the case that these 'pathogens of poverty' – expansively defined as physical, psychological and social, but largely focusing on the triad of sickness, spousal abuse and alcoholism – must be at the core of any analysis of growth in evangelical membership (ibid.: 7). Chesnut's approach has been highly influential but also animatedly disputed, for reasons that are worth considering briefly.

Against authors such as Burdick and Chesnut, whose attention is caught by the hopes evangelicalism seems to inspire of overcoming very specific challenges and difficulties within this life, critics objected that it is excessively reductionist to adopt a perspective that 'pay[s] attention to what religion *does* for its adherents' (Smilde 2007: 46, emphasis in original) in preference to what it *means* for them. For example, Peter Cahn took issue with the implication that one's religion is used as a kind of toolkit with which to solve specific problems arising in life and can be changed if it proves unsuitable for the purpose. Instead, he argued, faith is a holistic experience affecting every corner of one's existence and is thus not easily switched just because of a particularly intractable issue (2003: 116, 2005). Christian Smith likewise put the case for an approach to religion that understands humans as 'moral, believing animals' rather than handymen looking to 'fix' particular difficulties (2003: 80). Their emphasis on taking the meaning of converts' experience seriously was a helpful counterbalance to the sometimes heavy analytical reliance on structural forces as the sole determinants of conversion; however, they perhaps failed to acknowledge the extent to which the resolution of practical problems permeated the narratives of at least some groups of converts (Smilde 2007: 9). Moreover, their critique that the 'problem-solving' approach characterized converts as rational actors making instrumental decisions within a religious 'marketplace' (Smith 2003) underestimates the extent to which meaning may be thoroughly intertwined with the resolution of problems. In an evangelical worldview that contemplates not just the resources and shifts in lifestyle that joining the new congregation involves, but active divine intervention in the world and on the self, positive changes in one's circumstances act as confirmation of the 'rightness' of the decision to move. This is perfectly illustrated by an extract from one of my fieldwork interviews with Asunta, a member of the Maranata church who farmed a small landholding in a peasant community with her husband. Asked why she had changed from Catholic to Maranata a few years previously, she responded emotionally:

> My husband used to drink, he used to drink all the time around me. . . . He'll leave it, sure, when he loves God our Father, that drunkenness from before will be totally pulled out of him, I said. It's true, my husband used to be mad! Truly, he would hit me, he hit me so much, until I would lose myself, I'd run away. I used to go right into the cloud-forest when he would hit me. [Here she breaks off, sobbing.] Then when this happened to me like this, I believe totally, now he doesn't hit me anymore, and even if he drinks he doesn't hit me. He only drinks, he doesn't hit me. My husband used to stomp on my chest, he would even stomp! From that I'm injured, injured now. Even now! I'm not healthy anymore, not anymore. With those things [that he did to me] I, y'know – I bowed my head and approached Father God. Because of that he doesn't hit me anymore.

Asunta unhesitatingly and directly responded to my question about her conversion with the words 'my husband used to drink'. She attributed the change in his behaviour to God's intervention. For her, belief developed alongside the changes she saw in this area of her life, not because of a rigorous intellectual examination of the meaning of her new faith. Over the course of my fieldwork, it became apparent that many converts shared similar experiences and attributed to them the same importance in their decisions to change religion.

Another key question that has been asked of Chesnut's theory is why, if crises of illness or poverty are the sole or main determinant of conversion, whole populations have not converted en masse, for certainly in most Latin American countries such crises sooner or later affect the lives of the overwhelming majority (Smilde 2007: 9). Partial answers to this question come from David Smilde's own study of social networks and the extent to which they interact with meaning-centred factors in facilitating or discouraging those contemplating conversion (ibid.), and from the extent to which potential converts view the demands placed on them by their new religion as acceptable or even possible to meet (e.g. Manglos 2010). Yet it is increasingly recognized that the question itself contains unwarranted assumptions about how definitive conversion should be, for by no means everyone that joins a new church stays 'converted', or maintains active involvement with the church once they have done so. Moreover, many travel only a certain distance along the path to conversion before turning back. David Martin has referred to this as the 'revolving door' (2002: 112) of evangelical expansion. Scholars now more commonly speak of 'conversion careers' (Gooren, cited in Steigenga and Cleary 2007: 6) to capture this dynamic sense of partial moves, recidivism, wavering and disaffiliation. Such uncertainty and experimentation is particularly likely when studying contemplated conversions prompted – at least in part – by illness crises. While the unwell in my fieldsite did not change churches with anything like the abandon of, say, the Gwembe Tonga in Zambia (Kirsch 2004), neither were they as closely bound to either old or new denominations as their priests and pastors might have wished, and questions of efficacy remained key for them in deciding with which denominations they should align themselves.

The Fieldsite

Fieldwork took place in Paucartambo, a rural province of the southern Peruvian department of Cuzco. Most of its residents are peasant farmers; potatoes, barley, maize and quinoa are cultivated for sale and subsistence, while sheep, cows, llamas, hens or guinea pigs are raised to supplement income or diet. Most live in a myriad of dispersed villages known as *comunidades campesinas*,[3] which cling to the valley-sides and high moorland slopes. Roads, electricity and access to biomedical care expanded significantly, though not yet comprehensively, across these settlements during the 1990s and 2000s. A limited amount of wage work is available in the province's small administrative centres, but most non-farming income is gained from seasonal work either in the nearby jungle areas or from employment in Cuzco, the regional capital. Local men are almost universally bilingual in Quechua (Qu.) and Spanish (Sp.), while the majority of women over the age of thirty are to all intents and purposes monolingual in Quechua.

Though Paucartambo is famous for its sense of tradition and Catholic devotion, primarily because of its annual *fiesta* honouring the Virgin, evangelical Protestantism has nevertheless made significant inroads into the province. Agrarian Reform in 1969 abolished the quasi-feudal *hacienda* system that had acted as one brake on proselytizing efforts, and the way was thereby partly opened for missionaries to initiate projects of conversion. This met with scattered success initially, but gained pace in the 1990s so that today some villages are majority evangelical and barely a handful are still without resident Protestant families (Allen 2008: 43–45). For most families exposure to evangelicalism is still recent enough that one can speak of 'conversion' without qualms – as I do in this chapter – but some in the younger generation are now growing up without ever having self-identified as Catholic.

The largest evangelical denomination in Paucartambo by some distance is the Maranata church, which takes its name from the Aramaic expression meaning 'May God come!' According to Dominique Motte (2001: 71), it has been present in and around the Cuzco region since 1978, following rapid expansion from its initial base in the jungle. The Maranata gained many converts in Paucartambo towards the end of the twentieth century (Allen 2002: 223). Originally founded in Sweden on the principle of the 'priesthood of all believers', they remain one of the more tolerant evangelical groups (ibid.), but – in common with all evangelical churches in the area – nevertheless strongly prohibit the consumption of alcohol. Other churches with a significant presence in Paucartambo include the Evangelical Church of Peru, Adventist, Baptist and Assembly of God denominations. It is hard to generalize about converts to these churches; contrary to some predictions, there is no particular evidence that in Paucartambo evangelicals are more entrepreneurial or 'modern' in outlook than their Catholic counterparts, and classic markers of religious identity are no longer quite as salient as they once were. Abstention from alcohol is rightly seen as characteristic of evangelical identity, as I discuss below, but today many Catholics are also largely abstemious, in some cases surely following the Protestant example. In many villages, the traditional Catholic *cargo* system of sponsorship of communal fiestas has crumbled, not just because of the defection of resident evangelical converts from among the participants, but because of the adoption by numerous Catholics of what might be termed a 'modern' outlook, which deplores the expenditure and drunkenness that tended to mark traditional celebrations. In short, stereotypical markers of Catholic and Protestant identity fail to capture fully the complexities of contemporary religious allegiance in Paucartambo. In the remainder of this chapter, I explore the contribution of illness episodes to these shifting allegiances, what possibilities Paucartambinos perceive within evangelicalism as a repository of healing potential, and the implications for working with mental health issues in the province.

Navigating Catholicism and Evangelicalism in Southern Peru

During my fieldwork, I interviewed sufferers, their families, health workers, the local priests and pastors and a number of vernacular healers on their understandings of mental illness and approaches they might take to treatment.[4] One such interview was conducted with Antonia, a young villager afflicted with agonizing headaches and episodes of confusion, and her mother. Antonia described how she had once sought a cure from one of the local evangelical churches. Though she identified herself as Catholic, the possibility that the unfamiliar 'Saint-John-the-Baptists' might be able to improve her health had spurred her to investigate their practices. Nobody in particular had urged her to go, she said, but she had thought that they might have different powers to heal than the other local healers. Asked what they had done for her, she replied, 'Just the same as like Catholics – they made their supplications. . . . They pray, they supplicate – all that same old stuff.' Disappointed to find that the new religion seemed to have little more to offer her than the old, Antonia's deflated verdict was, 'They didn't make anything better.' Her mother, who disapproved of Antonia's former curiosity about the *hermanos*, was harsher: 'They made her worse.'

This exchange with Antonia is a reminder that prayer-based healing is not an altogether new phenomenon in the Andean context. While she was initially enticed by the apparent novelty of the evangelicals, she concluded that in many respects their therapeutic efforts resembled those of more familiar Catholicism. She may have had the formal Mass and the officiation of duly appointed priests in mind in saying this, but Catholic healing in these villages is more commonly accessed by petitioning a saint or the Virgin. This can be done through direct appeal to an image (usually backed up by a vow or offering), pilgrimage to particularly powerful local shrines or through the mediation of one of the many local lay intercessors whose prayers are reputed to have healing efficacy. Alongside these more orthodox approaches, healing can also be conferred by a range of spiritual forces inhering in the Andean landscape, most prominently the *apus*, or mountain spirits, that preside over much of human existence (Orr 2012, 2013). Such spiritual powers act through intermediaries, the vernacular healers generally known as *yachaqs* (Qu.: 'those who know'). Most locals perceive these entities to sit comfortably within an overall Catholic cosmology, reflecting the sedimentation processes of nearly five centuries of coexistence and entanglement. From the evangelical perspective, of course, this outlook adds paganism to the charges of which Catholicism is guilty, not so much for accepting the existence of these spirits as for appealing to them instead of, or as well as, to God.

Hence the evangelical offer of healing entered a field already saturated with religious frameworks for action in the face of illness. That some do not clearly differentiate it from this backdrop of other options is evident from comments such as this by the Catholic mother of a young man suffering from psychosis: 'I would have gone to the *hermanos*, but I had no money for an offering.' While full membership of the churches might indeed bring with it tithing obligations, and donations are sometimes sought and always welcomed, payment would not have been expected for praying for her son to be healed. Yet to enlist the assistance of saints, Virgins or other spirits, an offering of some sort would be required, in order to enter into the kind of relationship of reciprocity that might lead to the illness being cured. In this spirit, even public doctors told me how they were often presented with unsolicited gifts in attempts to establish reciprocity. This remains the understanding that underlies her statement about her decision not to seek healing from the Maranata church. The *hermanos* have clearly not managed to mark themselves off morally from other vernacular healers through their different approach to charging, as has been done by Pentecostals elsewhere (Pfeiffer 2006).

Against this, others – most notably converts themselves, unsurprisingly – take the view that evangelicalism has something dramatically new to offer in the face of traditional Catholicism. Here I quote a senior psychologist who worked at the principal specialist mental health institution in the region and was quite open about the influence her religious beliefs had on her work. Claiming to use only 'psychological techniques that are with God', she asserted:

> Culture is a medium with its own traditions and customs, and they imply rooted ideas which here are linked with Catholicism. The people of the *sierra* draw on these customs and make their children and grandchildren live a sad reality. These traditions make mental health suffer, because they cause adultery, fornication and substance abuse, and this produces 'revolving door patients' in those families which are immersed in tradition. They say 'I have to participate in my culture', not realizing that they are living in a sub-world, which makes the family and the society neurotic. Right now in Corpus Christi they come to the centre because of what they mistakenly call *fiestas*; it's the Roman Catholic religion that permits the deterioration of these people – because of it they become mentally ill automatically.

For her, evangelicalism offered the only feasible alternative to these pathogenic traditional religious forms. However, her openly evangelical stance was not typical of mental health professionals. Sometimes, the opposite viewpoint was apparent. For example, it is revealing that the 2006 Regional Mental Health Plan's analysis of threats to mental health listed two significant problems: 'lack of resources' and 'the sects' (DIRESA

2006). No discussion or guidance was provided on the precise scope of reference of the term 'sects', so it is not clear whether it was meant to include all non-Catholic churches, or only a subset of them. However, it is certainly significant that this is the terminology used by Pope John Paul II in a notable 1992 speech to the Conference of Latin American Bishops, in which he described evangelical groups as 'rapacious wolves' preying on the continent's Catholic communities (Cahn 2003: 69). In this light, the very ambiguity of reference is itself revealing of a commonly understood and shared meaning. In general, most mental health workers in and around Cuzco appear to give relatively little serious thought to religious distinctions among service users, but these two contrasting perspectives indicate the dramatic impact, both positive and negative, on mental health attributed to evangelicalism by locally influential figures.

At the community level, relations between the denominations are for the most part cordial, reflecting the necessity of getting along together for non-coreligionists living in small, often relatively remote, communities. There are occasional flashpoints of friction, but these usually concern prominent figures rather than the 'rank and file' of the faithful and seldom lead to more serious conflict. One such example occurred during an outbreak of what was claimed to be cholera some years ago, when the Maranata congregation sought criminal sanctions against a particularly successful *yachaq*, whom they blamed for spreading the disease through unsanitary healing practices. By extension the blame also fell on the 'pagan' petitioners, presumed to be Catholics, who sought his services. For his part, one of the area's Catholic priests accused one of the evangelical churches of facilitating orgies (which he claimed they called 'a moment with God') under the cover of their religious services, with grave implications for the mental well-being of young members of the congregation. All but the most fanatical of their congregations, however, maintained cross-denominational links and friendships. Negative attitudes towards each other surfaced from time to time, as Catholics sometimes spoke of the hypocrisy of their evangelical counterparts ('he walks with a bottle hidden behind his Bible' was an allegation made against supposedly teetotal Maranata converts) or evangelicals sniped at the drunkenness and immorality of the Catholics. More often such attitudes emerged in the ubiquitous stories of natural disasters that killed large numbers of Catholics but spared the evangelicals – or killed evangelicals and spared Catholics, depending on the allegiance of the teller.

Interestingly, a variant on the tales of divine punishment emerged from a family interview I conducted with the mother, Quintina, and sisters of a young woman, Daniela, who was suffering from confusion, forgetfulness and somnolence. We touched on the fact that her cousin had also developed a delirium recently, leading Quintina to lament, 'Who knows why this is happening to us?!' Daniela's older sister had a ready response: 'Don't you remember? It's because you were with the *hermanos*!' Quintina had indeed dallied with the Adventists some time previously, before returning to the Catholic fold; her daughter considered that the family's misfortune with mental disorders was attributable to this. For her, and for others who thought similarly, such a betrayal of one's own church invited supernatural punishment, perhaps in the form of madness. Links between religion and mental health risks thus ran through many of my informants' narratives.

Religion, Alcohol and Healing

I have outlined how choosing to belong to an evangelical church may be perceived in the province either to promote good mental health or to endanger it. But when it comes to specific instances of mental illness, many people, including Catholics, wonder whether these churches might have the power to restore them to health. The Maranata pastor, in the face of high expectations from many of those who approached him, was decidedly moderate in his claims as to what could be expected by sufferers of illness from his church. When asked what role he thought the church could play in healing illness, he listed – apart from the positive power of prayer in general – helping those suffering from addiction to alcohol or drugs to change their path, driving out the demon in cases of possession and praying for the cure of madness. Citing Job 37:19 and Ecclesiastes 7:7, he affirmed that demonic madness was real, was caused by the devil and would be appropriate for his church to treat. However, he distinguished this from natural madness, which he said was proper to doctors and which prayers would only rarely cure. Likewise, the Baptist pastor expressed the view that prayer can achieve much, but it is often necessary to learn to accept what cannot be changed. It may be that the pastors were deliberately circumspect with me to avert criticism; accusations are sometimes levelled that by overemphasizing the potential benefits of faith healing and downplaying alternatives they endanger the health of the people they advise (Kamsteeg 1991: 200), and indeed the head Catholic priest for the province described the evangelical engagement in healing as 'sensationalism' and clearly found this aspect of their ministry disturbing. Yet it appears that while both pastors accept the existence of effective and charismatic faith healers in the world, they do not expect such miracles to be replicated on any kind of regular basis in their own jurisdictions.

Despite the pastors' moderation, the trope of possible divine healing remains strong among many peasants in Paucartambo, as the cases of Asunta and Antonia show. This undoubtedly stems partly from the long tradition of popular belief in Catholic healing powers,[5] perhaps augmented by the sheer novelty of evangelical and Pentecostal modes of worship in a landscape where the familiarity of Catholicism has become stale for some. But the most significant factor in the link between evangelicalism and healing appears to lie in the particular association made between the Protestant churches and successful resolutions to problems of alcoholism. The importance of sobriety in inspiring conversion has been frequently highlighted in the scholarly literature on Latin America (e.g. Spier 1993: 110; Swanson 1994; Chesnut 1997: 111; Smilde 2007: 62; Allen 2008), but this analysis is by no means confined to the academy; it is a commonplace among residents of Paucartambo that one of the reasons people convert is to get away from drink, as Asunta's words quoted above have already shown.

An excerpt from one of my interviews with Gregoria, an abandoned (Catholic) wife whose husband Fausto had been an alcoholic, is another example of the extent to which the evangelicals are now considered an option of first resort for those affected by a drunken relative. She told me that she had long wanted to go to the Maranata expressly so that Fausto would stop drinking so much, but complained: 'He was a Catholic. He didn't want to enter the *hermanos*. "Go to the *hermanos*, go to them", how many times did I tell him! He just didn't want to go. I wanted to join the *hermanos*, too. I'm going to join them, I used to say, and I would have done. He didn't want to, he never wanted to, not for anything in the world.' Following up such statements more systematically by asking a wider circle of peasants (n. = 104) who were not members of the congregation where they might turn if a family member was a problematic drinker, I found that just under three-quarters of those in my informal survey (n. = 74) suggested either the Maranata, Baptist or occasionally the Adventist, churches, or made general reference to the evangelicals. Even those (mostly Catholics) who commented that their success in encouraging sobriety was exaggerated still acknowledged that they were well known for trying to play an important part in curing alcoholism – something for which neither the Catholic Church nor biomedical services were often given credit.[6]

It is not always realized that these churches' reputation for being able to assuage alcohol addiction has implications well beyond this specific condition. This is because alcoholism is not necessarily seen in Paucartambo as a simple state that stems solely from the physically observable and easily identifiable proximate cause of drinking. Long before Western biomedicine officially sought to reclassify alcoholism as a medical rather than moral disorder, it was already recognized by Andean ethnomedicine that its nature is contested and intricately bound up with other considerations relating to the overall physical, emotional and mental health of the person. There are perhaps three interpretations commonly placed on drinking to excess that are relevant here. The first is the perspective of moral condemnation – the drinker as someone dominated by *vicio* (Sp.: 'vice'), who is therefore seen as dissolute, selfish and a liability (see Harvey 1994). The second, meanwhile, focuses on the traditional importance of alcohol in cementing social bonds, not just with other humans but also with the spiritual world (Swanson 1994; Allen 2002: 124–26). The realm of the animate earth and the *apus* can potentially be accessed through inebriation (ibid.; Harvey 1994; Magny 2008); during the colonial period, realization of this association between drunkenness and pre-Christian rites led the Catholic Church to attempt to prohibit the Indians from heavy drinking during their festivities (Morales 2012). Seen in this historical light, the evangelical repudiation of alcohol is over-determined in the Andean context, due to the liquid's role in facilitating interaction with non-Biblical spirits. That this role holds true to the present day can be seen in the wry comment to me by Baltazar, one of the province's most successful *yachaqs*: 'If my patients don't see me drinking, they don't believe I'm healing.' The conflict that this sets up between evangelical and Catholic attitudes to alcohol is suggestive, implying that if evangelical believers are able to spurn such an important prop in forging spiritual alliances, they may have alternative means of accessing sources of power.

The third interpretation, however, is even more key to my argument here, as it makes a specific link between alcoholism, madness and other illnesses by tying together their causes – and hence suggests that success in improving the first may translate into success in curing the others. While drinking is connected to both moral failings and to supernatural power, it may also be viewed as a manifestation of illness (Harvey 1994: 227–28). Where drunkenness tips over into alcoholism it is often thought in this part of the Andes to be intricately bound up with soul loss, usually referred to in the Latin Americanist literature by the Spanish term *susto*, or in Quechua, *mancharisqa*. Usually occurring as the result of a sudden fright, which causes the soul, or *ánimo*, to start out of the body, this condition can have widely varying consequences for the affected individual, of which a propensity to alcohol abuse is only one possibility. There is a high degree of mutability in how *mancharisqa* may manifest in this province (Greenway 1998: 995; Orr 2012: 527), as a result of which it may be seen as either the sole or partial cause of a huge range of symptoms. The core symptoms, as elsewhere in Latin America, are usually listed as insomnia, lethargy, loss of appetite, feelings of restlessness and diarrhoea, but I have also heard epileptic fits, stammering, frenzied behaviour, an exaggerated startle reaction, tearfulness, one eye being sunken and half-open, emotional flatness, disordered reasoning, forgetfulness, headaches, talking to one's self, disgust for certain foods, delirium, fever, chronic pain, progressive paralysis, feelings of revulsion towards one's husband and addiction to alcohol attributed to it.

To illustrate how *mancharisqa* may emerge in local accounts, I recount the following story that Gregoria, whose ideas of getting help from the *hermanos* were reportedly frustrated by Fausto, told me about the origins of her husband's illness:

> His sickness took him. . . . It used to happen like an attack, he would fall to the floor. He was no good with his drink. . . . Opposite here once, when he was still a boy, an eagle came near him as he was running to get home. . . . And so it came down on him, like this [making a sudden swooping gesture] it came down on his head, and so he was wearing his *ch'ullu* [the knitted hat commonly worn by Andean men], and it took his *ch'ullu*, scratching him as it did. He was half-stunned. Juaniquillo[7] had got him, the devil Juaniquillo. . . . Baltazar Paucar told us that we would have to 'separate'[8] him. . . . I got Baltazar to call his soul. . . . And so I got him to call, but no! It was entering then, he was already nearly better – and just then he has a drink, just that! It was returning and just then it took him just like an attack.

The eagle's (Juaniquillo's) demonic apparition and attack had long ago caused Fausto the kind of shock that might be expected to result in soul loss. As Gregoria tells it, it was the long-term effects of this *mancharisqa* that later led to Fausto developing the need to drink habitually, hence the importance of a soul-calling ceremony orchestrated by a *yachaq* to allow him to overcome his addiction. Yet even as his *ánimo* was in the process of returning to him, it seems that Fausto was unable to resist the temptation to have a drink. This relapse undid the good work of the ceremony and the *ánimo* absented itself once more.

Fausto's current state of health appears from this narrative to stem from a complex interrelation between the act of drinking itself and the ongoing effects of soul loss. Although his addiction is said to have developed as a result of this loss, the action of continuing to drink seems to have had the effect of preventing the *ánimo*'s return and the anticipated recovery. There is thus allowance made for agency and the role played by an individual's own decisions, but the most important element in Fausto's downfall was the ongoing lack of the *ánimo*, which his wife argues affected his health and state of mind dramatically. The significance of this is that it shows that a causative link between this condition and drinking problems is one that makes sense to people. While of course there were – as one might expect – at least as many peasants who attributed the fault to Fausto's own character flaws and regarded the story of the loss of his *ch'ullu* as a convenient excuse, my point here is that the association is a real one for many, even if they may express cynicism in particular cases. Indeed, *susto* very often overlaps with other explanations of illness, seen as intermingling and mutually reinforcing. This makes intuitive sense if we consider that *mancharisqa* develops consequent to a fright; it is entirely reasonable that the cause of that fright might often be an injury or other scare related to the sufferer's health, which then both exacerbates and is exacerbated by the condition of soul loss. A good example of this is the case of Benigna, a woman in late middle-age who is also known for her uncontrolled drinking. According to her daughter, she became an alcoholic after suffering a dangerous fall that left her partially lame. On the one hand, she suggests that this is her mother's way of relieving the frustration of her now-limited mobility and the pain of her injured leg; on the other, she thinks that her mother may also be suffering from soul loss as a result of the shock suffered when she fell, and that this feeds into her current behaviour. There is no contradiction; rather the two states are bound up with each other.

It is unsurprising, then, given how easily *mancharisqa* can coexist with a variety of conditions of ill-health, that it may be invoked as an explanatory or contributory factor in a wide range of afflictions, from alcoholism to fevers, from psychosis to paralysis. To accept that the evangelical churches are often effective in managing alcohol addiction, as many in Paucartambo do, is in this context to acknowledge that they are able to ameliorate problems that are ultimately rooted in the loss of one's *ánimo*. Since so many apparently different forms of illness can be associated with soul loss, it follows that the *hermanos* may be able to resolve other illnesses in the same way. Contested as causation invariably is in surroundings where people are able to draw almost interchangeably on magical, religious, biomedical and folk notions of etiology (Orr 2012), the centrality and flexibility of the *mancharisqa* concept – while by no means accepted by all – at the very least entitle the evangelical churches to the benefit of the doubt over their effectiveness in healing afflictions that, as often as not, are thought to be rooted in precisely this phenomenon. Hence some seek evangelical help for such conditions as pulmonary disorders or heart conditions, and many consider the evangelical option when faced with serious mental disorder in those close to them. Relatively few of them, however, remain as lasting converts unless they experience a detectable degree of improvement or observe it in another family member.

Conclusion

I have shown how evangelicalism has become a significant social actor amidst the diverse ranks of healers working in the Peruvian Andes. Its therapeutic reputation has been hugely bolstered by the close associations made between its churches and the management of alcohol addiction. Although pastors may not actively promote the church as a healing panacea, often preferring to emphasize the less direct benefits of prayer and comfort, it nevertheless fits neatly into established expectations of religious institutions and of the nature of illness. The meaning or content of the evangelical message may yet have an important part to play for converts after a period of sustained involvement, at which point it is sometimes reinforced by success in facilitating at least partial recovery. However, the evidence suggests – *pace* Smith and Cahn – that 'instrumental' analyses of how at least some Andeans come to contemplate conversion, with a focus on attempts to resolve practical problems centring around illness and healing, do accurately describe decision-making processes in which they engage. Though the erosion of Catholicism's monopoly on Paucartambo's religious loyalties provides a new arena in which to make those choices, this is far from a radical departure from previous ways of relating to religion, for both the popular Catholic and distinctively Andean aspects of established cosmologies have always provided ways of understanding or acting on one's health when it is jeopardized. Hence the Maranata, Adventists, Baptists and others present new options to which Paucartambinos can turn for healing, but follow well-established precedents in doing so.

This aspect of my analysis belongs to a class of anthropological argument that has come under a degree of critical fire lately within the discipline. Joel Robbins has attacked the preponderance of what he calls 'continuity thinking' (2007: 5 and passim) as a framework for the discipline's approach to the study of Christian conversion. The accusation is that, because of a marked tendency to focus on cultural continuity, anthropologists have consistently been too ready to assume that conversion to Christianity is inauthentic or superficial. We are thus prone to see the traces or deep structures of earlier beliefs underlying whatever our interlocutors may in fact be telling us about their Christian beliefs and about the meaning of conversion, which – so the argument goes – is characteristically about 'rupture' rather than continuity. A secondary implication of Robbins's contention is that this readiness to dismiss what our converts are telling us in favour of what they 'really' mean is what predisposes many anthropological accounts to the view that their real reasons for the shift are 'everyday, pragmatic reasons – in search of things like money and power' (ibid.: 12) – and, it might be inferred, health.

Robbins makes a good case for his argument in his analysis of how anthropologists habitually devalue Christian discourse, while placing emphasis on pre-Christian elements of discourse and practice (ibid.). Before concluding, then, it is important to address this potential criticism of my analysis and, in the process, to clarify the extent and intention of my claims. First, I would suggest that 'rupture', while evident in the narratives of many converts in Paucartambo, is perhaps not quite so dominant as it was for the Urapmin with whom Robbins worked. While the Urapmin converted to Christianity within living memory, the last five centuries in the Andes have seen a process of continuous Christianization (under Catholic auspices) that repeatedly focused attention on what it is to be a 'real' Christian (Harris 2006). Consequently, while denominational transmigration is undeniably a shift, it is not as stark a break as the one described by Robbins (2004). Secondly, my intention is not to suggest that my informants remain 'Catholic' or 'syncretists' underneath – or conversely, that those who return to Catholicism after experimenting with the powers of the evangelical church are free-floating opportunists with no real loyalty to their religion. Rather, I seek to move away from the cognitivist bias that seeks to separate 'religious meaning' from 'pragmatic changes' in one's life. For most peasants in Paucartambo, religion is seen in immanent terms, with an expectation that it is relevant to the 'here and now' as much as it might focus on salvation and the hereafter (cf. Harris, in Robbins 2007: 22). Since the two are not easily separated in this worldview, there is no contradiction involved in exploring ultimate meaning through the practical effects of healing, whether that exploration takes place with the *hermanos* or through Catholicism. Thirdly, it is important to note that while healing, sobriety or both motivate a significant number of those who explore evangelicalism in Paucartambo, by no means all are impelled solely by these factors. My ethnographic focus was primarily on questions of mental illness and health, a research project that precluded full-scale investigation of conversion narratives across all families belonging to the evangelical churches in the province, so my intention is to call attention to an important trend, not to claim that it accounts for all conversions. I therefore do not exclude the possibility that 'rupture' talk may be more prominent in what other *hermanos* say. This brings me neatly to my fourth point: my analysis is firmly rooted in what my informants said to me. Healing was so prominent in their accounts that it was impossible to miss it. While comparison with Catholic and *yachaq* healing precedents is admittedly my own contextualization, there was no way that I could have ignored just how central healing possibilities frequently were in sparking interest in evangelicalism, nor how closely evangelicalism was linked with healing locally. Moreover, efficacy clearly mattered to my interviewees; those who had given up on the evangelicals did so, they said, because their health issues had not improved.

While thus acknowledging that there are risks in this approach to analysis, I would argue that much is also gained. The framework provides a perspective on healing among the different resources available in Paucartambo, helping to understand the relevance of the churches and patterns of resort commonly found among the rural Andean populace. In the role it plays in decision-making about religious loyalties, healing occupies a contested place in struggles for power, reflected in the harshly critical positions sometimes adopted by different figures of authority, both among senior mental health professionals and among church leaders. As a result of this, but also of the more general indifference among mental health workers towards the role of religion in the healing itineraries of their patients,[9] opportunities to address the failure to integrate mental health service input with families in the community are being missed. There are a number of factors influencing the high incidence of abandonment of mental health treatment in Paucartambo (Orr 2012); one is the lack of links between mental health services and rural communities at a distance from Cuzco. At the time of fieldwork, little had been done to address this; primary care was not notably successful at maintaining links with families caring for someone with a mental disorder. Given the frequency with which such families resort to the evangelical churches, and the openness of the pastors to recognizing that biomedicine might offer answers that prayer might not, some degree of collaboration between mental health services and the churches could provide useful options for signposting, monitoring and communicating key messages on mental health and the possibilities for input. Moreover, it would help understanding of the pathways along which families seek help, affording greater visibility to areas of patients' lives of which mental health professionals are often quite unaware.

The sick and their carers, in Paucartambo as elsewhere, engage with a variety of individual and institutional healing resources in their efforts to restore health, rendering once-firm aspects of their identities fluid as they do so. Religious flexibility is tied into this process; new and different churches loom large alongside the clinic and the *yachaqs* as a potential source of succour. Success in one area of suffering (alcohol dependency) becomes transferable, so to speak, to others through the mediator of *mancharisqa*, or soul loss, and informs the expectations that impel sufferers to go beyond their routine patterns of worship and explore new practices and discourses. In the Andean region, and through much of indigenous Latin

America, analysts of religious conversion and mental health specialists alike would do well not to ignore these processes and their effects on denominational membership, healing itineraries, and the subjectivities of those who make such choices about their spiritual, mental and physical well-being.

David M. R. Orr is a medical anthropologist and lecturer in social work at the University of Sussex. He carried out ethnographic fieldwork on madness in Quechua-speaking peasant communities in highland Peru, which has fed into his current research interests in the fast-developing field of global mental health. He is also working on a project in the United Kingdom studying service-user self-neglect and the complex dilemmas it poses to health and social care professionals in finding ways to navigate competing imperatives of autonomy and care.

Acknowledgements

This study was funded by the UK Economic and Social Research Council. Well-deserved thanks are also due to Roland Littlewood, David Napier, Penny Harvey and Simon Dein for their advice on the development of this chapter. I am especially grateful to my informants in Paucartambo, particularly the priests and pastors who welcomed me into their churches, and Grimanesa Toledo, whose help was invaluable in gathering the data I have presented.

Notes

1. Indeed, the important 2003 collection *Medical Pluralism in the Andes* (Koss-Chioino, Leatherman and Greenway), though it had much to say on the subject of health concerns, ethnicity, and their interactions with traditional Andean beliefs and Catholicism, barely even mentioned evangelical Christianity.
2. See reviews by Martin (2002) and Steigenga and Cleary (2007).
3. Literally 'peasant communities', which in Peru is a political designation entailing recognition of a communal identity that may affect the alienability of land rights.
4. The main period of fieldwork took place in 2007–8, with shorter visits in 2006 and 2010. Formal interviews were carried out with the family members of twenty-four individuals with mental disorders, eight mental health professionals, primary care medical staff and the Maranata and Baptist pastors. I attended a number of Maranata services in different villages, as well as a few Baptist services and the Catholic Mass. I observed healing rituals and was able to interview a total of fifteen healers. Extended participant observation in the village communities also enabled a multitude of informal conversations about religion, conversion, illness, madness and other relevant topics, from which I gleaned a wealth of information.

5. Despite the priest's scepticism about 'sensationalist' healing, many of his congregation continue to attribute miraculous powers to shrines, vernacular healers drawing on Catholic cosmology and material objects associated with the Mass.
6. Although the provincial health centre has run many state-sponsored campaigns to highlight the negative effects of alcohol abuse, it generally lacks resources, effective techniques or enthusiasm for tackling the problem in individuals.
7. Juaniquillo is a sinister spirit, identified by some with the devil. Others in the region thought of him as an entity in his own right, closely identified with the mestizo landowning class. Further discussion of local perceptions of Juaniquillo can be found in Greenway 2003: 95.
8. 'To separate' in this context is an expression for reclaiming a soul being held by spiritual beings.
9. Undoubtedly this partly stems from a feeling that delving in any depth into patients' religious beliefs is inappropriate territory for a mental health professional to be entering. There is something to be said for this view, as it helps to prevent any sense that the professional might be disrespecting the patient's beliefs or seeking to impose their own. Yet it also obscures considerable relevant information about the patient's life.

References

Allen, C. J. 2002. *The Hold Life Has: Coca and Cultural Identity in an Andean Community*, 2nd edition. Washington, DC: Smithsonian Institution Press.

———. 2008. '"Let's Drink Together, My Dear!": Persistent Ceremonies in a Changing Community.' In J. Jennings and B. J. Bowser (eds), *Drink, Power, and Society in the Andes*. Gainsville: University of Florida Press, pp. 28–48.

Burdick, J. 1993. *Looking for God in Brazil: The Progressive Catholic Church in Urban Brazil's Religious Arena*. Berkeley: University of California Press.

Cahn, P. S. 2003. *All Religions are Good in Tzintzuntzan: Evangelicals in Catholic Mexico*. Austin: University of Texas Press.

———. 2005. 'A Standoffish Priest and Sticky Catholics: Questioning the Religious Marketplace in Tzintzuntzan, Mexico.' *Journal of Latin American Anthropology* 10(1): 1–26.

Canessa, A. 2000. 'Contesting Hybridity: Evangelistas and Kataristas in Highland Bolivia.' *Journal of Latin American Studies* 32(1): 115–44.

Chesnut, A. 1997. *Born Again in Brazil: The Pentecostal Boom and the Pathogens of Poverty*. New Brunswick, NJ: Rutgers University Press.

de la Piedra, M. T. 2010. 'Religious and Self-generated Quechua Literacy Practices in the Peruvian Andes.' *International Journal of Bilingual Education and Bilingualism* 13(1): 99–113.

DIRESA. 2006. *Plan Regional de Salud Mental: Cusco*. Cusco: DIRESA.

Gamarra, J. 2000. 'Conflict, Post-conflict and Religion: Andean Responses to New Religious Movements.' *Journal of Southern African Studies* 26(2): 271–87.

Greenway, C. 1998. 'Hungry Earth and Vengeful Stars: Soul Loss and Identity in the Peruvian Andes.' *Social Science and Medicine* 47(8): 993–1004.

———. 2003. 'Healing Soul Loss: The Negotiation of Identity in Peru.' In J. D. Koss-Chioino, T. Leatherman and C. Greenway (eds), *Medical Pluralism in the Andes*. London: Routledge, pp. 92–106.

Harris, O. 2006. 'The Eternal Return of Conversion: Christianity as Contested Domain in Highland Bolivia.' In F. Cannell (ed.), *The Anthropology of Christianity*. Durham, NC: Duke University Press, pp. 51–76.
Harvey, P. 1994. 'Gender, Community and Confrontation: Power Relations in Drunkenness in Ocongate (southern Peru).' In M. McDonald (ed.), *Gender, Drink and Drugs*. Oxford: Berg, pp. 209–33.
Kamsteeg, F. 1991. 'Pentecostal Healing and Power: A Peruvian Case.' In A. Droogers, G. Huizer and H. Siebers (eds), *Popular Power in Latin American Religions*. Saarbrücken: Verlag Breitenbach, pp. 196–218.
———. 1993. 'The Message and the People – the Different Meanings of a Pentecostal Evangelist Campaign. A Case from Southern Peru.' In S. Rostas and A. Droogers (eds), *The Popular Use of Popular Religion in Latin America*. Amsterdam: CEDLA, pp. 127–44.
Kirsch, T. G. 2004. 'Restaging the Will to Believe: Religious Pluralism, Anti-syncretism, and the Problem of Belief.' *American Anthropologist* 106(4): 699–709.
Koss-Chioino, J. D., T. Leatherman, and C. Greenway (eds). 2003. *Medical Pluralism in the Andes*. London: Routledge.
Lazar, S. 2008. *El Alto, Rebel City: Self and Citizenship in Andean Bolivia*. Durham, NC: Duke University Press.
Magny, C. 2008. 'Quand on ne Peut Plus Boire d'Alcool ni Mâcher de Feuilles de Coca.' *Anthropology of Food, S4*. Available at http://aof.revues.org/index2972.html.
Manglos, N. D. 2010. 'Born Again in Balaka: Pentecostal versus Catholic Narratives of Religious Transformation in Rural Malawi.' *Sociology of Religion* 71(4): 409–31.
Martin, D. 2002. *Pentecostalism: The World Their Parish*. Oxford: Blackwell.
Morales, M. P. 2012. *Reading Inebriation in Early Colonial Peru*. Farnham: Ashgate.
Motte, D. 2001. *Una Revolución Silenciosa: El Impacto Social de las Nuevas Iglesias No Católicas del Perú*. Cuzco: CBC.
Muratorio, B. 1980. 'Protestantism and Capitalism Revisited in the Rural Highlands of Ecuador.' *Journal of Peasant Studies* 8(1): 37–60.
Olson, E. 2006. 'Development, Transnational Religion and the Power of Ideas in the High Provinces of Cusco, Peru.' *Environment and Planning A* 38: 885–902.
Orr, D. M. R. 2012. 'Patterns of Persistence amidst Medical Pluralism: Pathways toward Cure in the Southern Peruvian Andes.' *Medical Anthropology* 31(6): 514–30.
———. 2013. '"Now he Walks and Walks as if He Didn't Have a Home where He Could Eat": Food, Healing and Hunger in Quechua Narratives of Madness.' *Culture, Medicine, and Psychiatry* 37(4): 694–710.
Paerregaard, K. 1997. *Linking Separate Worlds: Urban Migrants and Rural Lives in Peru*. Oxford: Berg.
Pfeiffer, J. 2006. 'Money, Modernity, and Morality: Traditional Healing and the Expansion of the Holy Spirit in Mozambique.' In T. J. Luedke and H. G. West (eds), *Borders and Healers: Brokering Therapeutic Resources in Southeast Africa*. Bloomington: Indiana University Press, pp. 81–100.

Robbins, J. 2004. *Becoming Sinners: Christianity and Moral Torment in a Papua New Guinea Society*. Berkeley: University of California Press.
———. 2007. 'Continuity Thinking and the Problem of Christian Culture: Belief, Time, and the Anthropology of Christianity.' *Current Anthropology* 48(1): 5–38.
Scarritt, A. 2013. 'First the Revolutionary Culture: Innovations in Empowered Citizenship from Evangelical Highland Peru.' *Latin American Perspectives* 40(4): 101–20.
Seligmann, L. 1995. *Between Reform and Revolution: Political Struggles in the Peruvian Andes, 1969–1991*. Stanford, CA: Stanford University Press.
Smilde, D. 2007. *Reason to Believe: Cultural Agency in Latin American Evangelicalism*. Berkeley: University of California Press.
Smith, C. 2003. *Moral Believing Animals: Human Personhood and Culture*. Oxford: Oxford University Press.
Spier, F. 1993. 'Rural Protestantism in Southern Andean Peru: A Case Study.' In S. Rostas and A. Droogers (eds), *The Popular Use of Popular Religion in Latin America*. Amsterdam: CEDLA, pp. 109–25.
Steigenga, T. J., and E. L. Cleary. 2007. 'Understanding Conversion in the Americas.' In Steigenga and Cleary (eds), *Conversion of a Continent: Contemporary Religious Change in Latin America*. New Brunswick, NJ: Rutgers University Press, pp. 3–32.
Swanson, T. 1994. 'Refusing to Drink with the Mountains: Traditional Andean Meanings in Evangelical Practice.', In M. E. Marty and R. S. Appleby (eds), *Accounting for Fundamentalisms: The Dynamic Character of Movements*. Chicago: University of Chicago Press, pp. 79–98.
Theidon, K. 2014. *Intimate Enemies: Violence and Reconciliation in Peru*. Philadelphia: University of Pennsylvania Press.

Chapter 8

Cosmologies of Fear

The Medicalization of Anxiety in Contemporary Britain

Rebecca Lynch

Rates of anxiety within the United Kingdom have been found to be on the increase, with the UK government's Adult Psychiatric Morbidity Survey revealing that 9 per cent of individuals experience mixed depression and anxiety and 4.4 per cent generalized anxiety disorder (NHS IC 2009). Increasingly, these levels of anxiety have been attributed to living in more anxious times. Some have investigated what has made the times we live in more anxious and why we should respond to this in the way in which we do (Furedi 2007; Wilkinson 2001). These approaches view anxiety as a social 'problem', in opposition to the wealth of medical and psychological literature on the subject that views anxiety as an individual defect. The understanding of anxiety may be further developed, however, by taking an approach that links the social with the individual, relating this also to the medicalization of the experience of anxiety – through looking at anxiety, and the medicalization of anxiety, cosmologically. Within this approach, anxiety is viewed as a specific cultural response that relates to, and expresses, Euro-American cosmological ideas about the self and its existential relationship to the cosmos. This is influenced strongly by social and historical developments within Euro-American society, and subsequently results in the medicalization of anxiety as a cultural response to the experience. Such an approach attempts to further the understanding of anxiety as a Euro-American experience, but moreover, in placing anxiety in its cultural and historical context, may also

Notes for this chapter begin on page 172.

contribute to our understanding of cosmological notions and individual response within Britain and other Euro-American societies.

Research undertaken for the Mental Health Foundation (2009: 3) suggests that, as a nation, the United Kingdom is becoming more fearful; individuals perceive the world to be more frightening, and in turn feel more frightened. This survey found that 37 per cent of adults believed they get anxious or frightened more frequently than they used to, 77 per cent believed people in general are more anxious or frightened than they used to be and 77 per cent believed that in the last ten years the world has become a more frightening place (ibid.: 5). The survey reported that those interviewed were most anxious about the current financial situation (recession) (66 per cent), money/finances/debt (49 per cent), death of loved ones (45 per cent), crime or the threat of crime (35 per cent), the welfare of their children (34 per cent), developing a serious illness or disease (33 per cent), getting old (27 per cent), the state of the environment (18 per cent) and the threat of war (14 per cent) (ibid.: 21). Such concerns reflect the social environment and period of time in which the survey was conducted and reveal how fears may be embedded within a social context. A cycle of fear and risk aversion was also found by this survey, with perceived fear leading to risk aversion leading to actual fear (ibid.: 33). This report therefore links the increase in fear within the United Kingdom to increased numbers of individuals with clinical anxiety; if fear levels in the general public are high then more people will experience mental illness, and particularly the most common mental illnesses: depression, anxiety and anxiety disorders (ibid.: 1).

If levels of fear are related to levels of medical anxiety in the United Kingdom, medical anxiety would appear to reflect social and cultural aspects of living in the United Kingdom today. Furthermore, the examination of anxiety and its medicalization within the United Kingdom provides an example of how social experience and social distress may be taken on and experienced within the individual as personal distress, *and* how this is then dealt with culturally through medicalization of this individual experience. Such investigation into the cultural experience of anxiety also begs the question of explanations for the levels of anxiety and fear within the United Kingdom. It is particularly interesting that levels of anxiety have risen despite individuals living in statistically safer times than previous generations – for example, the fear of crime is still rising despite the fact that levels of crime have decreased in the last decade (ibid.: 3). What are we then talking about when we speak about fear and anxiety in the United Kingdom? How might these experiences be situated within the British context?

The Nature of Anxiety

Anxiety, as we know, is part of everyday experience. However, anxiety is also viewed as a clinical problem; the experience of anxiety is culturally transformed into symptoms when a particular point on the anxiety continuum is reached. The point at which clinical anxiety is sectioned off from 'normal' anxiety is difficult to define but for some authors, this is where anxiety is obstructive in day-to-day life and where clinical intervention would be beneficial (Gale and Davidson 2007). A range of explanations and approaches have attempted to understand (and treat) anxiety of a clinical level. The evolutionist idea that anxiety results from a 'fight-or-flight' response has been picked up by some psychologists such as Michelle Craske, who, like other psychologists, views anxiety as a product of maladaptive thinking, resulting from inappropriate upbringing and socialization. She notes that individuals with generalized anxiety disorder (GAD), for example, continually detect and interpret possible threats, overestimating the probability of these threats and seeing themselves as ineffective at managing them. A cycle is then created; a negative personal view is then seen as evidence of individual ineffectiveness, leading to increased pessimism (Craske 2003). Increasingly popular psychological treatments fit with this idea, including techniques such as cognitive behavioural therapy (CBT) that aim to address the problem of anxiety through altering this maladaptive thinking process within the individual so that they are again able to operate within society. Freud also wrote much about anxiety, with his views on what he termed 'anxiety neurosis' changing over his lifetime, from a reaction to trauma to a problem of transformed libido (1993 [1925]). These ideas, and the treatment of anxiety through psychoanalysis, have influenced psychiatric understanding of anxiety, although pharmacological interventions are also offered for anxiety disorders and reflect biomedical ideas of anxiety as having an underlying biological cause (Trimble 1996; Rees, Lipsedge and Ball 1997). Psychological, biological and psychoanalytic ideas mix within psychiatry, but all such approaches focus on anxiety within the clinically 'ill' individual, rather than on wider contributing causes, as the result of individual deficit. This is in stark contrast to sociocultural approaches that focus on anxiety as problematic in wider society rather than located solely within the individual.

Sociocultural approaches to the problem of anxiety remind us that anxiety and fear are socially constructed and culturally conditioned responses, aspects to anxiety that are less visible in medical conceptualizations. What we fear and the strength of that fear depend on conceptions of the world, the perilous forces that reside within it and our options for protection against these (Svendsen 2008: 24). While cross-cultural approaches to emotion vary (Milton and Svašek 2005), few anthropologists would contest that our cultural perspective constructs what we view as fearful or anxiety-promoting, and that this same cultural view may then promote the extent and expression of this fear and anxiety. The cultural and social impact on fear is also suggested by the temporal quality of many fears. Both Lars Svendsen and the historian Joanna Bourke suggest that all time periods have their fears but that *what* is feared changes over time (Svendsen 2008; Bourke 2005). This is clearly visible in the films and literature of science fiction within the United Kingdom over changing time periods; Mary Shelley's Frankenstein was created at the time of, and arguably reflected the fears of, the industrial revolution, for example, while a proliferation of books about nuclear war emerged during the period of the Cold War (Susan Sontag's 1965 essay 'The Imagination of Disaster' also deals with changing fears expressed in Japanese and American science fiction films). Within Euro-American culture today, what Svendsen terms a 'low intensity fear' (2008: 46) or 'constant weak "grumbling"' (ibid.: 76) exists as the dominant form of fear and provides a background to our experience and the way in which we interpret the world. Consequently this 'grumbling' can be seen as more of a mood than an emotion (ibid.: 46), and this culture of fear is emblematic of our period in time and a metaphor through which we view our experiences (Furedi 2007). In line with Frank Furedi, Svendsen contends that Euro-American cultures consider nearly all phenomena from a perspective of fear despite living in a more secure position than ever before (2008: 7).

Although intimately related, anxiety can be conceptually separated from fear in relation to personal experience. Both Bourke (2005) and Svendsen (2008) suggest that fear refers to an immediate, 'objective' threat. Anxiety, however, is an anticipated 'subjective' threat; anxiety comes from within and is more generalized (Bourke 2005), lacking a specific object with a nature of 'indefiniteness' (Svendsen 2008: 35) – the 'constant weak grumbling' Svendsen mentions above rather than a direct threat. Anxiety is therefore 'deep', whereas fear is 'shallow' (ibid.: 9). Furthermore, in fear, individuals are able to assess the situation and neutralize or flee the problem; however, those subjectively experiencing anxiety are unable to act (Bourke 2005). For the sociologist Iain Wilkinson anxiety leaves individuals searching for cultural forms that adequately express the true origins and identity of the anxiety (2001). Wilkinson proposes that where individuals remain entwined by anxiety therefore, culture has not provided a means by which the feeling of being overwhelmed by the uncertainty of the future can be dealt with (ibid.: 131). Wilkinson and Furedi thus link the growth of anxiety and fear to modernity and the growth of the 'risk society' (Beck 1992) within Euro-American culture.

As well as the factors that might create anxiety changing over time and being connected to cultural and social circumstances, how anxiety is expressed may also differ cross-culturally. Responses to distress differ across cultures and the expression of anxiety that has been medicalized by biomedicine might be seen as a particularly Euro-American presentation. Research on the many examples of what were formerly termed 'culture-bound syndromes' demonstrates different expressions of distress located particularly within different societies. This term has largely been dropped due to its suggestion of a restrictive, fixed and bounded nature of such expressions and research has come to view these presentations as culturally specific collections of symptoms and culturally constituted means of displaying distress, or 'idioms of distress' (Nichter 1981). Roland Littlewood further suggests these might be seen as 'stylized expressive traditional behaviours' that have moderately similar presentation, can be time-limited and, while going against everyday 'normal' behaviour, are condoned within the culture as an expression of distress (2002). Such idioms may not necessarily be pathologized by these cultures however. Work on *'ataque de nervios'*, a cultural expression of distress found in Puerto Rico (and in Spanish-Caribbean individuals elsewhere) describes *ataque* as 'an experience accessible to certain groups when bad things happen' that is understood to result from a cultural context in which a gender-based expectation of social control exists (Lewis-Fernández et al. 2009). *Ataque* can be seen an expression of distress best understood within its cultural context. Such a response can be compared to the 'laments' described by James Wilce in Bangladesh where individuals express their distress through wept singing, a very different form of expressing distress but one in which Wilce argues individual identities can be constructed and resistance to power expressed (1998). Both laments and *ataque* are culturally condoned forms that may be used to express the anxieties encountered in life. Although there is not space here for a full discussion of the 'symptoms' of Euro-American anxiety, placing anxiety in Euro-American cultures in the context of high modernity, where the self is viewed in an alternative cosmological way to the past, anxiety symptoms perhaps express physically these Euro-American notions of self and disconnectedness as well as other Euro-American cultural notions of the body itself.

The dominance of Euro-American medicine and of medical categories has made cultural differences peculiar to these cultures easy to miss. Given that the majority of these categories were first described in Euro-American cultures based on their own populations, Euro-American idioms of distress have been incorporated into such definitions, and held up as a standard form, from which other cultural expressions deviate in exotic fashions. Placing the medical categories of psychiatry in their cultural and historical context highlights the Euro-American cultural specificity of these labels however, not only of the 'pathologies of the West' such as anorexia nervosa and multiple personality disorder described by Littlewood (2002), but also more common mental health problems such as depression (Kleinman and Good 1985; Skultans 1979; Showalter 1987). The symptoms of Euro-American anxiety can be seen as culturally sanctioned responses, idioms of distress that 'make sense' culturally – they are a cultural response both in the sense that they respond to sociocultural circumstances *and* in the cultural patterning of how that anxiety is expressed. For Euro-American cultures to therefore medicalize anxiety, to construct a diagnosis and label these experiences as in need of medical treatment, is a further cultural response to these cultural responses, indicative of the status of biomedicine and how experiences become incorporated into the medical sphere. Euro-American culture has dealt with this increased anxiety through medicalization and it is to the medicalization of anxiety that I now turn.

The Medicalization of Anxiety

Littlewood suggests that in Euro-American cultures, in general, distress is medicalized: it is 'seen through a lens which encourages us to experience and indeed shape, individual concerns in medical ways'; the illness comes from outside, with a cause, pattern and perhaps a cure (2002: 1). Anxiety also fits this notion; the symptoms of anxiety are culturally recognized as a 'medical' problem with individuals seeking assistance from primary care general practitioners (family doctors) rather than the priests who would have been more commonly consulted for anxiety in previous times (Bourke 2005).

The connection between healing and religion are evident in many cultures and were once more explicitly linked within Euro-American societies. Religious orders previously took a central role in caring for the sick, however modernity promoted a split between the religious and medical domains, with improvements in science developing the medical understanding of the human body and its treatment, and power moving from the unproven and unquestioning belief of religion (already unsettled through the Reformation) to the demonstrable evidence and rational thinking of science during the Enlightenment. In addition to the movement away from religious personnel to medical personnel in care and treatment for the sick, wider concerns around suffering and salvation, previously the domain of the church – concerns that Byron Good terms 'soteriological issues' (1994) – also moved to the domain of medicine. Good argues that moral and soteriological issues are 'fused' with medical issues and that medicine mediates the 'physiological' and the 'soteriological', illness having both physical and existential dimensions as it reveals the infirmity of the body and human suffering. For Good, cultures are organized around a soteriological view through which the nature of suffering is understood and salvation is achieved. In Euro-American cultures medicine is 'the core of our soteriological vision' (ibid.: 70), perhaps also reflecting the reduced power and influence of the church in Euro-American culture. Additionally, however, as argued above, these 'soteriological issues' – how we suffer, live, die, make sense of life – are the very stuff of Euro-American anxiety, perhaps making anxiety particularly susceptible to being placed under the frame of medicine.

Bruno Latour and Steve Woolgar tell us that scientific 'facts' are socially constructed (1979), but furthermore that such facts and medical labels are also culturally produced. Earlier anthropological work by authors such as Allan Young (1995) and Littlewood (2002) have considered the medicalization of experience into illness, illness categories in Euro-American culture and the effects and consequences for those involved. Through such work, experience is seen to be translated into 'symptoms' and behaviour into 'pathology', the 'abnormal' sectioned off from everyday experience into the pathological realm. Books such as Young's work on PTSD have illustrated how medical labels may be brought about for particular social or political purposes, such as the construction of the category of railway spine for insurance claimants (Young 1995) as well as the use of and creation of medical labels by drug companies to develop new pharmaceutical markets (Lakoff 2008; Watters 2011). Such labels can also legitimize behaviour and create or reduce positions of power, and Bourke claims that fear itself sorts individuals into hierarchical social positions. She gives the example of 'school-phobia' being used for middle-class children (with working- class children given the label of 'truancy') and states that fear – but we may also add medical labels around fear and anxiety – can be related to the distribution of power (2005).

The power exerted by medicine, both through the power of medicalization and by the moralizing aspect of medicine, is further illustrated in Elaine Showalter's consideration of 'hysteria' in the British context. She suggests that medicine took a moralistic role in controlling female sexuality and actions, with suffragettes and 'modern' women desiring to work or divorce their husbands among those diagnosed with 'hysteria', rendering them as 'mad' and in need of (medical) control. Medical management is therefore a way of containing women's suffering without confronting its causes (1987), and Furedi notes the reorientation of social problems into individual emotional problems in contemporary 'therapy culture' (2004). The great advantage of this relocation is, of course, that it is the problem within the individual that becomes the focus of treatment, and wider social issues do not need to be addressed – and this may be particularly the case for a society such as the United Kingdom, where there has been a strong drive towards an individual responsibility for health. Moralizing aspects of medicine can also be found in recent debates around individual responsibility for health and obesity, smoking, and the promotion of 'health behaviours' to support a 'healthy lifestyle'. These prescriptions on how to live again may be seen as reminiscent of the moralizing discourses and position of moral guides previously held by the Christian Church in earlier history. In adding to other discussions on medicalization that focus on the medical system and its positioning and power to medicalize, in this chapter I take a slightly different perspective on the medicalization of anxiety. I situate anxiety in its cultural and historical context and consider how changes to concepts of self and the role of the church impacted on both experiences of anxiety and the development of medicine. This is therefore a discussion of medicalization that focuses on the cosmological and how cosmological concerns have become viewed as part of the realm of medicine.

Cosmological Approaches to Anxiety

Rather than focusing specifically on arguments purely around medicine therefore, I turn my attention to understandings of Euro-American (and, in particular, British) cultural changes over time and how cosmological understandings of the world, the individual within it and perceptions of risk and control might have led not only to increased anxiety but also to its medicalization. Sociological and broader social theory approaches to anxiety view levels of anxiety found in (Euro-American) society as largely resulting from the current period of late/high modernity (Wilkinson 2001; Giddens 1991) and it is this relationship that I want to go on to examine in more depth.

The relationship of modernity to notions of risk is seen as fundamental to this largely sociological approach. Both Anthony Giddens (1991) and Ulrich Beck (1992) have considered risk itself as central to late modernity, 'fundamental to the way both actors and technical specialists organize the social world' (Giddens 1991: 3). This suggests that modernity produces a 'risk society' or 'risk culture', where public knowledge and debates about risk and the riskiness of everyday life, as well as the introduction of new types of risk previous generations have not faced (such as nuclear war and environmental breakdown), are present in the everyday life of the individual. Despite the actual overall reduction in life-threatening events for the individual (Lupton 1999), this greater knowledge of risk, Giddens and Beck argue, increase our insecurity about our individual position in society and our ability to live in safety and make us more anxious about the future ahead of us.

In addition to greater awareness of perceived risk and the related insecurity this brings, risk society is also about not only what *has* happened, but what *could* happen, and where no one is out of danger (Svendsen 2008: 48–50). In fact, Pat Caplan argues that whereas previously the past was used to determine the present, the future, as this is seen through various risk scenarios, is now used to determine the present, with history being of little significance (2000). Such a focus on risk and future risk is also, by its very nature, related to uncertainty and attempts to control. Svendsen suggests that uncertainty is a basic element of human life and argues that (Euro-American culture) is dominated by the 'precautionary principle' as a response to dealing with such uncertainty (2008: 67). Such a principle constructs a world where the future is made up of dangers rather than possibilities (ibid.: 71). Svendsen cites the *British Medical Journal*'s decision to forbid the use of the word 'accident' in its pages as symptomatic of the view that the world is completely controllable (ibid.: 64). It is also worth noting here that this prohibition occurs in a medical journal; not only is the world viewed as controllable but medicine in particular is portrayed as fully understanding the world. Science now guides the individual rather than religion (Furedi 2007; Svendsen 2008; Bourke 2005) and authors such as Giddens (1991) argue that the process of modernity decreased the role of the church and promoted a scientific and rational worldview. But as the historian Bourke notes, science has dispelled superstitions while also delivering new fears and new risks (2005), including risks beyond individual human control such as those related to the environment and the political sphere.

How dangers are conceptualized and dealt with has changed through modernity, not only through notions of risk but also through increased individualization. Individualism was seen to rise at the same time as the growth of risk (within modernity) and with it what Caplan calls an 'ongoing search for morality' (2000: 6), where individuals look to control and improve themselves rather than the social environment. In fact, the very position of the individual in relation to the world around them has changed through modernity, with the individual, rather than external forces, located as the seat of power. As suggested by Deborah Lupton, individuals in Euro-American cultures feel they maintain a high level of control over danger and their exposure to it; risk is therefore viewed as the responsibility of humans rather than through notions of fate or destiny (Lupton 1999), and therefore the church plays no role. Individuals thus bear greater responsibility, but, stripped of the support and guidance previously afforded by the church, are also more vulnerable and isolated to deal with the risks of which they are increasingly aware. Risks are not only practical and related to the social environment but also relate to the security of the individual, producing existential anxiety.

Taking these understandings of risk society and modernity further, anxiety can be interpreted as existential angst, a situation where individuals struggle to create meaning in their lives in the face of the uncertain world that surrounds them (May 1996). This idea has also been considered by Anthony Giddens, who links anxiety and the notion of 'ontological insecurity' in relation to the process of modernity. The concept of ontological insecurity was developed previously by R. D. Laing, who saw the ontologically secure person as one who is able to meet life's problems 'from a centrally firm sense of his own and other people's reality and identity' (1990 [1960]: 39). Ontological security gives the individual the experience of self and the ability to relate to the world around them, a world that the individual organizes through what Giddens terms 'basic trust' (1991: 38). This trust, developed through the individual's upbringing and socialization, is connected to the individual's identity and is the 'protective cocoon' that they carry with them to be able to continue with the activities of everyday life (ibid.: 40). It is argued therefore that the process of modernity contributes to feelings of ontological *in*security, which in turn brings about feelings of anxiety. This anxiety also creates further feelings of insecurity by impeding the awareness of a sense of self as it challenges the confidence of the relationship between self and outside world (ibid.).

The precursor, and an explanation for the historical development of 'ontological insecurity' in Western culture, is found in the argument of Erich Fromm in his work *Escape from Freedom*.[1] Fromm takes an existential psychological position that all humans have a need to feel related or connected to the outside world; humans have a need to avoid isolation and therefore individual freedom from the bounds tying the individual to the world are important. Religion and other belief systems give protection from 'aloneness' and security, without which life lacks meaning (Fromm 1969). For Fromm, individuals in Euro-American cultures have become 'more free' (including existentially free), but without the religious hold that previously gave security. This process started with the breakdown of mediaeval feudal society, which now gave individuals freedom from previous economic and political ties but also freedom from the ties that provided a sense of belonging and security. Before this breakdown, Fromm argues, there was no notion of the individual 'self'. People were part of a family, a village; the construction of the universe was simple and the relationship with God based on confidence and love. Afterwards, the individual was free but anxious and alone, seeing others as potential competitors, existentially threatened and with a view of God that was also less secure. Luther's theology gave expression to this experience and offered a solution: moving away from church authority to an individual relationship with God. But in so doing, the individual needed to accept their own insignificance and powerlessness, and leave behind the notion that humans had salvation, and instead view life as being about economic productivity. Protestantism therefore helped the individual deal with their anxiety, reorientating the individual to this novel world and developing an individualistic worldview as ties from others were lost and the individual faced God on their own. Weber, despite having an otherwise similar view of Protestantism, saw a novel cultural spirit and economic behaviour as developing over time (1992 [1930]), while Fromm suggests that society moulded a particular social character within the individual, which formed the basis of new cultural ideas (1994).

For Fromm then, the individual self and its connection to society became uncertain and thrown into doubt through historic changes in the mediaeval period, but the anxiety this generated was coped with through religious attachment and the drive to work. Such ideas are a useful grounding for theories of modernity that have considered the changing role of the church after this period, the increased secularism and individualism pursued through the process of modernity, and the increased vulnerability of the individual self. For example, Wilkinson suggests that anxiety may be a modern term for an age-old feeling, part of the very nature and psychology of being human (2001: 45). In such a secular culture, he suggests, psychological language has replaced religious language in the way we explain our experiences:

> When it comes to discussing matters of feeling, we now prefer to speak with deference to the authority of experts in the fields of human science rather than the (more doubtful) wisdom of those who would explain our problem in terms of our relationship toward God. However in taking up the language of anxiety, perhaps we have not only come to explain ourselves differently, but further, we may also have begun to modify the way we feel. (Ibid.: 45)

Anxiety can thus be linked to cultural experiences of modernity, not only through the process of modernity resulting in greater awareness of risk, individualization and ontological insecurity, but also through increased secularization. The diminishing role of the church, as well as cultural changes that altered understandings of self, others and the divine, meant that those experiences of everyday life that were formally the domain of churches became increasingly the realm of medicine and psychology. These changes suggest cosmological shifts in how the world was understood to operate; cosmological issues moving from being questions for churches to deal with to questions for medicine and psychology.

The Psychologization of Self and Ontological Insecurity

Problems of the self are not only discussed in psychological language in Euro-American cultures, but are also psychologized; Horacio Fabrega, for example, suggests that mental illnesses themselves are 'disturbances of the self' (1992: 100). The self, placed within the individual body through individualism, is thus expressed through the body, and it is interesting to note that one of the 'symptoms' of anxiety, and a number of other mental illnesses, is a feeling of 'depersonalization'; the individual feels detached from the self and an outside observer of what they are doing and thinking (American Psychiatric Association 2000). The relationship between anxiety and a sense of self can also be seen in the historical use of lobotomies to treat anxiety and fear; Bourke notes that these procedures were 'successful', as they destroyed a sense of self (2005). Changes over time in the view of the self and the move from religious assistance to psychological help has also been noted by Bourke in her historical review; whereas in the past the anxious individual might have turned to the church for comfort, as the self was located within the body through the twentieth century, anxiety was more individualized and treated through therapy or self-help (2005). For Furedi, this therapy has had a strong influence in Euro-American culture – a 'cultural phenomenon rather than a clinical technique' (2004: 22) – which demonstrates Euro-American culture's new focus on the importance of emotion and how this domain is dealt with (ibid.).

Furedi argues that within Euro-American 'therapy culture', therapy is linked to identity, a 'therapeutic script' used to understand the self. Furthermore, therapy culture has cultivated emotional vulnerability through the endorsement of a position where the self is seen as limited and fragile, in need of ongoing therapeutic intervention, and without emotional resources to cope with adverse circumstances – a vulnerable self (ibid.). Interestingly, Furedi also points out that theologians are now 'therapists' (ibid.), to which the growing literature of the field of pastoral psychology is testament. This also represents an opportunity for religious figures to re-engage with individuals suffering from anxiety, albeit using techniques and approaches developed from more clinical interventions, which therefore keep anxiety within a medicalized framework.

As Fromm notes, historical social changes within Western culture have placed the site of control within the individual self. This is reinforced by the scientific worldview, for example through the psychological notion of the 'locus of control',[2] the extent to which the individual feels personal control of their actions and the world around them. Having a high internal locus of control is viewed as more positive than having a high external locus of control, with the latter previously seen as more prevalent in poorer populations and thus an explanation for poorer health in these areas (Poortinga, Dunstan and Fone 2007). This individual locus of control situated in the self is related by Littlewood to experiences such as anxiety. He suggests that anxiety, as well as other expressions and emotions, communicate a representation of the self that has lost self-will and -control, a loss that can be short-term or permanent, partial or complete (2002: 185). The content of such experiences – the framework, expectations and responses around these – organize individual narratives that also illustrate what Littlewood terms 'the experiential reality of our local cosmology' (ibid.: 186). Through the process of modernity, therefore, there has been a psychologization of self, and the site of control has been situated within the individual; the self is discussed in psychological language, disturbances of the self have been pathologized and self-control has come to be seen as a psychological trait.

Research within medicine and psychology have also examined, and found a place for, ontological security (including through religious faith) in alleviating anxiety, research that also demonstrates how these concepts have themselves become part of psychology. Aaron Antonovsky's (1979, 1987) notion of our 'sense of coherence' (SOC), the way in which human beings make sense of the world and use the required resources to respond to it, can be viewed as strikingly similar to the notions of ontological security and cosmological understandings. Studies that have investigated SOC have found a negative correlation with anxiety and depression and a positive correlation with optimism and self-esteem (Hart, Hittner and Paras 1991). Some psychiatrists and psychologists recognize that spirituality and religiosity emphasize the depth of meaning and purpose in life and that religion is a coping strategy for dealing with life events (Dein et al. 2010). They note that many religions also hold a 'just world hypothesis' (Hogg, Adelman and Blagg 2010); good things happen to good people, bad things happen to bad people, and in this way the world has method, consistency and purpose. Furthermore a 2009 study found that when conducting tasks measuring uncertainty, participants with greater religiosity and a stronger belief in God had a reduced reaction in the cortical system involved in self-regulation of anxiety. The authors concluded that religious conviction [and the ontological security it is connected to], act as a buffer against anxiety and provide a framework for action and understanding of the environment (Inzlicht, McGregor, Hirsh and Nash 2009). Anxiety and ontological insecurity have therefore also become part of medical and psychological studies, but as these studies demonstrate, so has religious faith itself. Attempts to understand religious faith from within medicine have focused on psychological benefits, such as security and support, which emerge from religious faith; an approach which then rationalizes the religious within a scientific worldview. Rather than being thought of as 'ignorance' of scientific 'knowledge'[3], religiosity and religious experiences are, from this perspective, thought of as fulfilling a psychological need (or symptomatic of psychiatric disorder) but resulting in the religious domain remaining subordinate to the medical field.

Within the medical field, psychiatry itself has been seen as lower status and less 'medical' than other specialties, psychiatry and mental health services typically receiving less funding and focus than other areas. Psychiatry has been viewed, and is still viewed by those within medicine and outside, as less scientific than other medical disciplines. Dealing as it does with illnesses that are often less obviously attributed to a solely biological base and without biological tests (both of great importance within a medical worldview), the psychiatrist works more prominently in an uncertain world – another strike against it from the certainty-loving scientists. The need to make psychiatry more allied to science and thus more 'medical' and therefore more distant from religion, may mean that psychiatrists in particular psychologize cosmological understandings of anxiety, ontological insecurity and religion more broadly, focusing on the need for a medical 'cure'. This relates to research findings where not only does a substantial difference exist between the religiosity of the American population and American psychiatrists (psychiatrists being far less religious) (Lukoff, Lu and Turner 1992), but also psychiatrists are found less likely to be religious in general than other medical disciplines (Curlin et al. 2007). These findings link to ongoing debates about the role of religion in psychiatry and the 'religiosity gap' between psychiatrists (and other mental health clinicians) and mental health patients (Dein, Cook, Powell and Eagger 2010; Lukoff, Lu and Turner 1992). In Farr Curlin et al.'s study, religious physicians were also found to be less willing than nonreligious physicians to refer patients to psychiatrists (2007). David Lukoff, Francis Lu and Robert Turner (1997) note the historical tensions between religion and psychiatry, which they attribute to the close links between psychiatry and psychoanalysis, and therefore to Freud's anti-religious stance (1992). The ability to 'explain away' religion as a coping method by psychology and psychiatry may also be added as a possible contributor, as demonstrated in Michael Hogg, Janice Adelman and Robert Blagg's notion of the 'Uncertainty-Identity Theory' to 'account' for religiousness.[4]

Religiosity and cosmological ideas have over time therefore become part of psychology, transformed into testable and quantifiable concepts to be measured scientifically. As noted above, science has given Euro-American cultures a worldview based on certainty, a cosmology founded on rationality. This view also privileges the power of 'evidence'. Through a scientific worldview, a god cannot exist as there is no [scientific] 'evidence', as Richard Dawkins's book *The God Delusion* tells us (2006). Evidence replaces and devalues experience, taking it out of the personal realm to be treated as a separate, and often measurable, 'thing' of its own, a disease with the possibility of 'cure'. Anxiety therefore can move from being an experience to being a diagnosis. Its changing nature and cultural embeddedness is stripped away through this process as its components and its diagnostic label should be discrete and objective, able to be applied to any body in any place at any time so that treatment can be instigated. The foggy and unstable boundaries of anxiety have, however, caused problems for the classification of anxiety, and in this last section I focus on some of the ways the field of medicine has tried to deal this.

Medicine Deals with Anxiety

Once biomedicine had staked a claim to anxiety, it then had to find a place for it within wider medical categories. The scientific approach of defining illness categories, creating order from disorder, is particularly difficult in the case of anxiety. Like many other mental illnesses, definitive medical 'tests' to ascertain the presence or not of the disorder are unavailable and, as previously noted, the key areas about which individuals worry change over time and can depend on the individual as well as social circumstances – e.g. economic recession bringing fears related to finance and financial stress (Mental Health Foundation 2009). How these are interpreted in a scientific framework, in which Latour and Woolgar suggest 'social' factors 'disappear once a fact is established' (1979: 23), is challenging and may contribute to the changing descriptions of medical anxiety over time.

To try to deal with some of the range of presentation and concerns within the broad category of anxiety, biomedicine has created subcategories within the broader classification of 'anxiety', different symptoms indicating different subgroups of the disease and suggesting different treatment plans. However, separating these different groups, and distinguishing anxiety from other conditions such as depression, has not been straightforward. The National Institute for Health and Clinical Excellence (NICE) guidelines produced for clinical practitioners in the United Kingdom notes the difficulty in distinguishing the anxiety disorder 'sub-types' from each other, and indeed some differences in categorization exist between the categories given by NICE, the contemporary version of the World Health Organization's International Classification of Disorders (ICD-10) and the contemporary revision of the American Diagnostic and Statistical Manual of Mental Disorders (DSM-5) (National Institute for Health and Clinical Excellence 2004; World Health Organization 1992; American Psychiatric Association 2013). These differences relate largely to the emphasis placed on different symptoms and the number of symptoms needed to be present to make a diagnosis (Gale and Davidson 2007), and illustrate that this category is not immutable even by medical standards.

Before 1980 those with severe feelings of anxiety would have been diagnosed with 'anxiety neurosis' (Barlow and Wincze 1998), a condition first described by Freud in 1895 (1993 [1925]). This connection to Freud and his method of psychoanalysis remained linked to the condition of anxiety until the publication of the third edition of the Diagnostic and Statistical Manual of Mental Disorders (DSM-III), which introduced the separation of anxiety disorders under the wider category of anxiety. This also brought anxiety more firmly into the realm of medicine, and clinical care of anxiety was initiated by research that suggested a more medical basis to one of the anxiety disorders. The anthropologist Byron Good, who advised on the development of the DSM-III, notes the excitement around the category of panic disorder at this time (2002), as recently published articles had found that patients with panic disorder responded to anti-depressants, unlike patients with generalized anxiety disorder (GAD), and that panic attacks could appear 'unprovoked . . . out of the blue'. These factors suggested a biological basis for what had previously been viewed as a 'psychological disturbance and the strong hold of psychoanalysis' (ibid.). In her book on the history (and personal experience) of panic disorder Jackie Orr suggests that the new DSM was based on empirical and observable symptoms, categorizations and diagnoses based on tests and techniques that 'became central to psychiatric thinking'. Through these changes, psychiatric researchers then overtook clinicians as the 'most powerful force in the profession'; in comparison, claims from psychoanalysis were viewed to be un-provable and less relevant (Orr 2006: 225). The historical (and cultural) situatedness of the DSM and the diagnostic categories it puts forward were further illustrated through the discussion of proposed amendments to the category of GAD in the fifth edition of the DSM (DSM-5) (Lewis-Fernàndez, personal communication, 2010). Again, such changes illustrate the difficulties in placing anxiety in a rigid medical framework.

Biomedical categories are of course created through research and over time and are not 'naturally' occurring. Latour and Woolgar remind us that scientists attempt to produce order, struggling to impose a framework that reduces 'background noise', giving an apparently logical and coherent outcome (1979: 36–37). Uncertainty is not welcomed by science or by biomedicine (as the banning of the word 'accident' in the *British Medical Journal* testifies) (Davis and Pless 2001) and part of the scientist's role is to create the order our worldview requires: 'order is the rule . . . disorder should be eliminated wherever possible' (Latour and Woolgar 1979:251). Through setting up categories, uncertainty can be lessoned as the world is set in order, as convincingly argued by Mary Douglas (2002 [1966]). This creation of order from disorder arguably was also previously the realm of formal religion and the church in separating the sacred from the secular and promoting a divine ordering through which the individual understands personal experience. The very basis of anxiety is uncertainty and disorder however, and therefore it is not surprising that historical changes to the dominant framework through which the world is ordered by society have resulted in changes to the social institution that deals with the problem of anxiety. Both institutions, the church and medicine, have wielded great power at different time periods to interpret individual experience (not least the experience of anxiety) in their own terms. With the reduced influence of the church, the growth of biomedicine and the fields of psychiatry and psychology together with changes to understandings of the cosmological, anxiety might be seen as particularly ripe for medicalization.

On these bases, how might the data on increased anxiety at the start of this chapter be understood? Are we actually becoming more anxious or might rates of clinical anxiety merely point to increased *diagnosis* of anxiety? More frequent diagnosis might also be attributed to changes in expressions of anxiety, clinical definitions, wider attitudes towards mental illness, knowledge of clinicians and even the actions of drug companies looking to act on new markets. Furthermore, can we assume that all existential angst has been medicalized, or that that the psychological and medical realm has full dominance over cosmological understandings? For many patients, for example, medicine and the scientific perspective do not explain everything. Medicine may explain why two individuals had heart attacks, but it cannot always explain why one died and the other did not; it may explain the how of a situation but not necessarily the why. These understandings can be linked to *umbaga*, the 'second spear' found in Evans-Pritchard's study of the Azande (1976 [1937]); while, in his famous example, it was understood that the granary fell because it was being eaten by termites, *umbaga* provides the explanation as to why it fell on those people at that particular time. These understandings can be found too in UK ethnographic work on 'lay' perspectives of illness, for example in Charlie Davison, Stephen Frankel and George Davey Smith's study on explanations for illness in a Welsh village (1992). Here, health promotion messages were counteracted by stories of 'Uncle Norman', who had smoked and drunk all his life only to die in his late nineties (while another individual who had been healthy all their life had died suddenly at a young age). Notions of 'luck' and 'fate' were used to complement more medical perspectives, explanations that have not been completely removed through the increasing power of scientism. These notions of 'luck' and 'fate' attribute cause to an external agency, perhaps not surprising given the tradition of viewing a God as 'up there', unlike traditions elsewhere such as the Yolmo, where cosmological ideas are represented in the individual body and in society as well as in the wider cosmos (Desjarlais 1992). This is not incompatible with ideas of individualism; Paul Heelas suggests that in cases of a strong emphasis on the autonomous self, deviations from what we wish to occur are attributed to a discrete agency that is external to the self (1981, cited in Littlewood 2002: 184). Perhaps, then, anxiety has become medicalized but is also not entirely resolved by medicalization; people's cosmological worlds are not entirely taken over by a medicalizing force, nor are they entirely passive agents to biomedicine's increasing dominance.

Through taking a cosmological perspective of anxiety, the reasons for its medicalization are perhaps not surprising given the increased role of science and medicine over the previously dominant religious structure that has resulted through the process of Euro-American modernity. Euro-American culture and the process of modernity not only contribute to increasing anxiety therefore, but are also involved in creating a particular type of Euro-American anxiety linked with Euro-American notions of the self and the cosmological position of the self in relation to the world around it. Culturally, again in response to changes stimulated by the process of modernity in these societies, anxiety is handed to clinicians as part of the domain of medicine. This is not to say, of course, that other cultures do not experience anxiety, as certainly they do, but that there is something *particularly* Euro-American about the pattern of distress expressed through clinical anxiety and how this is then dealt with through medicalization. Through this perspective, anxiety emerges as a cultural response profoundly affected by culturally specific actions and reactions and linked inexorably to the Western process of modernity in its construction, experience and resolution. I have sought to draw out how changing cosmological understandings might emerge from wider cultural shifts, and how these in turn might result in changing cultural responses.

This is not just a discussion about anxiety and medicalization, therefore, but seeks to contribute another means by which examination of the cosmological might reveal something about mental illness.

Rebecca Lynch is a research fellow in medical anthropology at the London School of Hygiene and Tropical Medicine (LSHTM). Her work crosses the intersection of medicine and religion and she is particularly interested in understandings of morality, health and the body and cultural and scientific constructions of these. She has conducted fieldwork in Trinidad and the United Kingdom.

Notes

1. *Escape from Fear* was published with the title 'Fear of Freedom' in the United Kingdom in 1941.
2. The psychological notion of 'locus of control' was developed from Rotter's 1954 concept of social learning theory, which suggests that expectations about the future are developed from previous experience (cited in Poortinga, Dunstan and Fone 2007).
3. Good (1994) makes the point that lay health understandings have typically been termed 'beliefs' and contrasted with scientific 'knowledge' to emphasize the validity of scientific vs. nonscientific understandings of the world.
4. Hogg, Adelman and Blagg's 'Uncertainty-Identity Theory' suggests that religions have attributes that make them well suited to reduce feelings of self-uncertainty, as individuals are able to lessen such feelings through identification with groups. While all groups provide belief systems and normative prescriptions, they argue, religions also address the nature of existence and provide a moral compass, making religious affiliation particularly attractive in uncertain times (2010). Such a theory is similar to Fromm's argument regarding the appeal of religion, both also accepting that self-uncertainty can be related to ontological certainty.

References

American Psychiatric Association. 2013. *Diagnostic and Statistical Manual of Mental Disorders: DSM-5*. Washington, DC: American Psychiatric Association.
Antonovsky, A. 1979. *Health, Stress and Coping* San Francisco: Jossey-Bass.
———. 1987. *Unravelling the Mystery of Health: How People Manage Stress and Stay Well*. San Francisco, CA: Jossey-Bass.
Barlow, D. H., and J. Wincze. 1998. 'DSM-IV and Beyond: What is Generalized Anxiety Disorder?' *Acta Psychiatraica Scandinavia* 98 (Suppl. 393): 23–29.
Beck, U. 1992. *Risk Society: Towards a New Modernity*. London: Sage.
Bourke, J. 2005. *Fear: A Cultural History*. London: Virago Press.
Caplan, P. 2000. 'Introduction.' In P. Caplan (ed.), *Risk Revisited*. London: Pluto Press, pp. 1–28.
Craske, M. G. 2003. *Origins of Phobias and Anxiety Disorders: Why More Women Than Men?* Amsterdam and London: Elsevier.

Curlin, F. A., et al. 2007. 'The Relationship between Psychiatry and Religion among U.S. Physicians.' *Psychiatric Services* 58: 1193–98.
Davis, R. M., and B. Pless. 2001. 'BMJ Bans "Accidents".' *British Medical Journal* 322: 1320–21.
Davison, C., S. Frankel and G. Davey Smith. 1992. 'The Limits of Lifestyle: Re-assessing 'Fatalism' in the Popular Culture of Illness Prevention.' *Social Science and Medicine* 34(6): 675–85.
Dawkins, R. 2006. *The God Delusion*. London: Black Swan.
Dein, S., et al. 2010. 'Religion, Spirituality and Mental Health.' *The Psychiatrist* 34(2): 63–64.
Desjarlais, R. R. 1992. *Body and Emotion: The Aesthetics of Illness and Healing in the Nepal Himalayas*. Philadelphia: University of Pennsylvania Press.
Douglas, M. 2002. *Purity and Danger: An Analysis of Concept of Pollution and Taboo*. London: Routledge.
Evans-Pritchard, E. E. 1976 [1936]. *Witchcraft, Oracles and Magic among the Azande*. Oxford: Clarendon Press.
Fabrega, H. 1992. 'The Role of Culture in a Theory of Psychiatric Illness.' *Social Science and Medicine* 35: 91–103.
Freud, S. 1993 [1933]. *New Introductory Lectures on Psychoanalysis*. London: Penguin Books.
Fromm, E. 1969 [1941]. *Escape from Freedom*. New York: Henry Holt and Company.
Furedi, F. 2004. *Therapy Culture: Cultivating Vulnerability in an Uncertain Age*. Abingdon: Routledge.
———. 2007. *Culture of Fear: Risk-taking and the Morality of Low Expectation*. Trowbridge: Cromwell Press Ltd.
Gale, C., and Davidson, O. 2007. 'Generalized Anxiety Disorder.' *British Medical Journal* 334: 579–81.
Giddens, A. 1991. *Modernity and Self-identity: Self and Society in the Late Modern Age*. Cambridge: Polity Press.
Good, B. 1994. 'How Medicine Constructs its Objects.' In Good, *Medicine, Rationality and Experience*. Cambridge: Cambridge University Press, pp. 65–87.
———. 2002. 'Culture and Panic Disorder: How Far Have We Come?' *Culture, Medicine and Psychiatry* 26: 133–36.
Hart, K. E., J. B. Hittner and K. C. Paras. 1991. 'Sense of Coherence, Trait Anxiety, and the Perceived Availability of Social Support.' *Journal of Research in Personality* 25(2): 137–45.
Hinton, D., and B. Good. 2009. *Culture and Panic Disorder*. Stanford, CA: Stanford University Press.
Hogg, M. A., J. R. Adelman and R. D. Blagg. 2010. 'Religion in the Face of Uncertainty: An Uncertainty-Identity Theory Account of Religiousness.' *Personality and Social Psychology Review* 14(1): 72–83.
Inzlicht, M., et al. 2009. 'Neural Markers of Religious Conviction.' *Psychological Science* 20(3): 385–92.
Kleinman, A., and B. Good. 1985. *Culture and Depression*. Berkeley: University of California Press.
Littlewood, R. 2002. *Pathologies of the West: An Anthropology of Mental Illness in Europe and America*. Ithaca, NY: Cornell University Press.

Lakoff, A. 2007. 'The Private Life of Numbers: Pharmaceutical Marketing in Post-Welfare Argentina.' In A. Ong and S. J. Collier (eds), *Global Assemblages: Technology, Politics, and Ethics as Anthropological Problems*. Oxford: Blackwell Publishing Ltd, pp. 194–213.

Laing, R. D. 1990. *The Divided Self: An Existential Study in Sanity and Madness*. Harmondsworth: Penguin Books.

Latour, B., and S. Woolgar. 1979. *Laboratory Life: The Social Construction of Scientific Facts*. Beverly Hills, CA, and London : Sage Publications.

Lewis-Fernández, R., et al. 2009. 'Comparative Phenonmenology of "Ataque de Nervious", Panic Attacks, and Panic Disorder.' In D. Hinton and B. Good (eds), *Culture and Panic Disorder*. Stanford, CA: Stanford University Press, pp.135–56.

Lukoff, D., F. Lu and R. Turner. 1992. 'Toward a More Culturally Sensitive DSM-IV.' *Journal of Nervous and Mental Disease* 180: 673–82.

Lupton, D. 1999. *Risk and Sociocultural Theory, New Directions and Perspectives*. Cambridge: Cambridge University Press.

May, R. 1996. *The Meaning of Anxiety*. Revised edition. W. W. Norton & Co.

Mental Health Foundation. 2009. *In the Face of Fear: How Fear and Anxiety affect our Health and Society, and What We Can Do About It*. Report prepared by the Mental Health Foundation, available at www.mentalhealth.org.uk.

Milton, K., and M. Svašek. 2005. *Mixed Emotions: Anthropological Studies of Feeling*. Oxford: Berg.

National Institute for Health and Clinical Excellence. 2004. *Anxiety: Management of Anxiety Panic Disorder, with or without Agoraphobia, and Generalized Anxiety disorder in Adults in Primary, Secondary and Community Care*. Clinical Guideline, available at guidance.nice.org.uk/CG22.

NHS IC. 2009. *Adult Psychiatric Morbidity Survey in England, 2007. Results of a Household Survey*. The NHS Information Centre for Health and Social Care. Available at www.ic.nhs.uk/pubs/.

Nichter, M. 1981. 'Idioms of Distress: Alternatives in the Expression of Psychosocial Distress: A Case Study from South India.' *Culture, Medicine and Psychiatry* 5: 379–408.

Orr, J. 2006. *Panic Diaries: A Genealogy of Panic Disorder*. Durham, NC, and London: Duke University Press.

Poortinga, W., F. Dunstan and D. Fone. 2007. 'Health Locus of Control Beliefs and Socio-economic Differences in Self-rated Health.' *Preventive Medicine* 46(4): 374–80.

Rees, L. R., M. Lipsedge and C. Ball. 1997. *Textbook of Psychiatry*. London: Arnold.

Showalter, E. 1987. *The Female Malady: Women, Madness, and English Culture, 1830–1980*. London: Virago Press.

Skultans, V. 1979. *English Madness: Ideas on Insanity, 1580–1890*. London: Routledge and Kegan Paul.

Svendsen, L. 2008. *A Philosophy of Fear*. London: Reaktion Books.

Trimble, M. R. 1996. *Biological Psychiatry*. 2nd edition. Chichester: John Wiley and Sons.

Watters, E. 2011 *Crazy Like Us: The Globalization of the Western Mind*. London: Robinson.

Weber, M. 1992. *The Protestant Ethic and the Spirit of Capitalism*. New York and London: Routledge.
Wilce, J. 1998. *Eloquence in Trouble*. Oxford: Oxford University Press.
Wilkinson, I. 2001. *Anxiety in a Risk Society*. London: Routledge.
World Health Organization. 1992. *The ICD-10 Classification of Mental and Behavioural Disorders: Clinical Descriptions and Diagnostic Guidelines*. Geneva: World Health Organization.
Young, A. 1997. *The Harmony of Illusions: Inventing Post-Traumatic Stress Disorder*. Princeton, NJ: Princeton University Press.

Chapter 9

Functionalists and Zombis

Sorcery as Spandrel and Social Rescue

Roland Littlewood

All social anthropologists are 'functionalists'. They are functionalists in that when they examine a society – or a segment of a society, or a cultural area – they generally see all the components, institutions, actions, concepts and mentalities as somehow fitting together. And each is only to be understood completely through examining its neighbouring components (what we might call 'weak functionalism'). As Rivers once put it, 'it is hopeless to expect to obtain a complete account of any one department without covering the whole field' (1914: 1). Anthropology textbooks tell our students that elements of a social formation are no longer to be dismissed as meaningless or as the survival of a once intelligible purpose. Even now when we pursue practice or precept across related communities we assume that (for us) these elements can generally be understood only as part of the wider picture (cf. Littlewood 2007). Indeed, wary of defining out 'medicine' or 'warfare' in a particular culture, we try to hold them against other institutions or patterns of interest, such as kinship and family organization, land use and gender, in our holding up to students our claim that we practice the most holistic of scientific enterprises.[1]

An ethnography without an element of 'function' – without recourse to an idiom of solidarity, of satisfying fears and dilemmas, resolving feuds, of channelling creativity, answering existential questions, or contributing to local cohesion – would be bare indeed.[2] Whether we take matters as far as our ancestors in the heyday of functionalism to proclaim that each little element necessarily has some vital value (Stocking 1984) in contributing to

Notes for this chapter begin on page 186.

the coherence or stability of the social whole (Radcliffe-Brownian 'strong' structural-functionalism) is less persuasive (Radcliffe-Brown 1977). The problem was of course not only in temporarily circumscribing areas of interest but in assuming that societies could be seen as self-contained, consistent and constant functioning wholes, whilst ignoring the imaginative justifications that anthropologists could eventually produce.[3] And certainly, few would now seek to follow the later Malinowski in isolating a single element and attempting to show how it directly subserves not only social cohesion but 'basic human needs' (in his psychological and biological 'strong functionalism') (Kaberry 1957).

It was in the area of witchcraft confessions and accusations that a particular functionalist model – the so-called homeostatic theory – became popular in the years following Evans-Pritchard's (1937) classic work on the Azande. Such accusations were assumed to provide a systemic feedback function for a society, preserving existing social relations. Collecting examples together, Mary Douglas (1970) criticized the crude homeostatic models of the 1950s to take into account that bugbear of the functionalist, social change, in arguing that witchcraft accusations happen, still she admitted, as an adjustment mechanism, but in particular situations of competition and ambiguity.[4] In the zero-sum economics of small-scale communities, sudden accession to wealth was a matter for suspicion of unfair advantage and thus demands for redress.

Edwin Ardener (1970), in that same volume, describes how in the precolonial period the introduction of cocoyam initiated a period of prosperity for the Cameroon Bakweri. Unequal wealth and the envy of others seem to have been resolved through potlatch-type ceremonies, agonistic prestations in which the more wealthy competitively feasted others and then destroyed their own possessions – a functional alternative to sorcery accusations. Subsequent German colonization was accompanied by the establishment of plantations using external labour, and thence a general decline in the Bakweri local economy, in self-confidence and morale, and with the growth of prostitution and venereal disease, along with an increase in accusations of sorcery – these took a new form in that the sorcerers now had power over their dead relatives, who became a labour force for the sorcerers (a recognition confirmed by Geschiere 1998) as zombie-like spirits. Later, in the 1950s, the Bakweri took up banana cultivation as a cooperative enterprise; general prosperity returned and the proceeds of banana production were used to fund a 'witch-cleansing' movement in which both the sorcerers and their zombies disappeared. There has more recently been a return of the Bakweri zombies (S. Ardener personal communication).

Summarizing Ardener's model: economic competition in a situation of general poverty where distinctive wealth is suspect is associated with increased accusations of sorcery, the zombies enabling the zombie masters to become relatively prosperous. That the sorcerers (or witches, as Ardener called them) are potential rivals of others within the same sphere (and more generally the European in Africa, like the Zande noble, the later Roman emperor and the early modern magistrate, could not be either witch or victim) alerts us not only to real relations in the economic world but to the psychology of actual or presumed invidia. Douglas notes that accusations are often a legitimate way of breaking off relations with those for whom there were expectations of mutual dependence.[5]

Whilst Haitian zombification is only confirmed when the escaped victim is found wandering, attempting to return to their grave, one cannot assume that the victim found and interviewed by us is necessarily the same physical personage as the one locally presumed to have been abducted and zombified. This is relevant when considering the Harvard Zombie Project of the 1980s, when NASA-funded pharmacologists and ethnobiologists sought a form of suspended animation for space flight, and wondered if zombification in Haiti was a distinct psychopharmacological process. This was presumed to involve neurotoxins such as tetradotoxin (from the puffer fish *Sphoeroides testudineus* and *Diodon hystrix*), with *Datura stramonium* (*concombre zombi*, zombie cucumber) used to revive and then control the zombi (Anon 1984; Davis 1988). Tetradotoxin had been studied biomedically in Japan where the puffer fish ('fuga fish') is a dangerous delicacy whose consumption may indeed result in apparent but often temporary death (Anon 1984).

Studies at that time of the one well-documented instance of a returned zombi (Davis 1988; Douyon 1980) concentrated on his symptoms at the time of presumed death, with little on his mental and physical state at the time of the post-return interview – although a lay observer (Thomson 1992) did not remark on any abnormality at this later time. In fact, few identified zombis have been clinically assessed and no clinical or experimental evidence has been published on the state of the surviving zombi until a 1997 paper in the *Lancet* medical journal (Littlewood and Douyon 1997). This current chapter takes up the earlier issue of the identification of the presumed zombi to speculate on any apparently independent, socially 'functional' purpose served by zombification.

Haitian Zombis

In spite of the similarity between the Bakweri word *sombi* (to pledge a relative to zombification) and the Haitian Creole *zombi*, Ardener argues that West Indian zombie slave concepts are 'a spontaneous growth' (1970: 148). Zombification seems to have come to international notice during the occupation of Haiti by the United States between 1915 and 1934 (Hurston 1938; Hurbon 1993). A late nineteenth-century account of Haitian belief and practice by the British consul Spencer St. John in 1884, whilst hardly neglecting macabre tales of human sacrifice and cannibalism, does not mention it, although he does refer to burial alive after administration of a narcotic. The US occupation seems to have resulted in greater local competition for limited resources; it permitted foreign ownership of land and the establishment of plantations and was met with a number of revolts.[6] Popular American military memoirs, and then Jacques Tourneur's 1942 film, *I Walked with a Zombie*, popularized the escaped zombie theme, as did Zora Neale Hurston's credulous but sympathetic book *Voodoo Gods* (1938), which offered the first photograph of a zombi (in the psychiatric hospital).

The current UN intervention has again focused international attention on a phenomenon regarded as exotic and improbable by Western public media, yet which is taken by most Haitians as a distinct and empirically verifiable state. Along with the related practice of *vodu*, it has been implausibly related by American physicians to the recent epidemic of AIDS in Haiti (Farmer 1992a). Haitian medical practitioners I spoke with regard zombification as the very real consequence of poisoning; the clergy accept it as the magical product of sorcery. Zombis are frequently recognized by the local population, and medical estimates of their number are of the order of up to a thousand new cases per year.[7]

Zombification is a crime under the Haitian Penal Code (Article 246), where it is considered as murder even though the zombified individual is still 'alive'. Local interpretation is that either by poisoning or sorcery (and the two are not easily distinguished, for *poison* can act at a distance), a young person suddenly and inexplicably becomes ill, is subsequently recognized by their family as dead, placed in a tomb, only to be stolen by the *bòkò* (sorcerer, often a male *vodu* priest or *oungan*) in the next few days, and then secretly returned to life and activity but not to full awareness and agency (Métraux 1958; Mars 1945). Haitians are seldom buried underground but rather interred in painted concrete tombs that, in country areas, are on family land next to the houses: generally secured with a not very robust metal plate or door, these tombs are certainly vulnerable to being broken open. The central cemetery in Port-au-Prince includes a large number of smashed and opened tombs whose contents have been removed.

A number of different local schemata of body, mind and spirit all recognize something like a separation of the *corps cadavre* (physical body) with its *gwo-bon anj* (loosely, animating principle) from the *ti-bon anj* (agency, awareness and memory) (cf. Davis 1988; Hurston 1938; Métraux 1958). The latter is retained by the sorcerer, usually in a fastened bottle or earthenware jar where it is known as the *zombi astral*; the *bòkò* either extracts it through sorcery, which leaves the victim apparently dead, or else captures it after a natural death before it has gone too far from the body. The animated body remains without will or agency as the *zombi cadavre*, which becomes the slave of the sorcerer and works secretly on his land or is sold to another for the same purpose (as a *zombi jardin*). It is induced to remain a slave only through chaining and beating, or through further poisoning and sorcery. This *zombi cadavre* is the zombie popularized by Western cinema and indeed is locally referred to simply as the *zombi*.[8] In Haiti, as in Britain, the term is also used in explicit metaphor to refer to extreme passivity and control by another.

Different explanations as to how a *zombi cadavre* may escape back to its original family suggest that either the bottle containing the *zombi astral* breaks, or the *bòkò* inadvertently feeds his *zombis cadavres* salt, or he dies and they are liberated by his family, or – rarely – they may be released through divine intervention. On release, now called a *zombi savan*, their mental and physical status remains the same, and they are vulnerable to recapture and continued enslavement. Few *bòkòs* or medical doctors claim to be able to return a *zombi cadavre* to its original state of health and agency, and the matter is reserved for the mercy of *Le Grand Maitre* (the rather remote God recognized by *vodu* practitioners who is only invoked briefly through Latin prayers when they begin their ceremonies). Zombis are regarded with commiseration – fear being reserved for the possibility of being zombified oneself.[9]

Concern that a deceased relative may be vulnerable to zombification justifies prevention through the not infrequent decapitation of the corpse before burial, or poisons and charms are placed in the coffin to deter resurrection, or an eyeless needle and thread are provided to distract the corpse endlessly.

An Instance

Wilfred D. was twenty-six years old in 1996, the eldest son of an alleged former *tonton macoute* under the Duvaliers (and who is still a significant figure in the hill village near Les Cayes where the family live). The father was the only senior member of the family who had attended elementary school, and he was my principal informant together with Wilfred's mother and other villagers. Wilfred had been intelligent and industrious, and a cousin in town had promised to pay for a college education if he did well at the secondary school he attended in the departmental capital.

When he was eighteen he suddenly became ill with a fever, 'his eyes turned yellow', he 'smelled bad like death' and 'his body swelled up'. Suspecting sorcery, his father asked his own elder brother to obtain advice from a *bòkò* while a female cousin prepared him some magical soup. However, Wilfred died after a three-day illness (in other accounts seven) and was buried in a tomb on family land next to the house of this cousin. Inexplicably, the tomb was not watched that night. Nineteen months later Wilfred reappeared at a nearby cock fight, was able to recognize his father, and according to the latter accused the uncle of having zombified him with the help of a *bòkò*; Wilfred had correctly recalled comments made by his family during his earlier funeral ('we'll keep his school uniform'). He was immediately recognized as a zombi by the other villagers, by the local Catholic priest and the magistrate, and has since remained in his father's house, his legs secured to a log to prevent him wandering away.

The uncle was arrested at the father's request and sentenced by the provincial court to life imprisonment for zombiflcation, confessing that he had indeed been jealous of his younger brother, who had used his literacy to register all the family land in his own name. This uncle also implicated the cousin on whose land Wilfred was buried; she was arrested but released after interrogation. According to Wilfred's friend who had first identified him in the crowd at the cock-pit, Wilfred had then mentioned having stolen two coconuts from his uncle's plot who, encountering him drinking them, seized them to use for sorcery. (Another friend gave a rather different account and said the jealous uncle had probably rubbed an astringent magical poison on Wilfred's arm.) The father's story was confirmed to me by the villagers, the judge and priest involved in the court case and by the local coffin maker, and by examination of Wilfred's death certificate and the proceedings of the trial. The uncle escaped from prison during the political turmoil of 1991, was traced and agreed to an interview in 1996 in exchange for armed protection; he denied sorcery or poisoning, saying that the recovered Wilfred was an impostor and the case was a trick on the part of Wilfred's father to expropriate him entirely, and that his own confession had been induced through torture by the local police. (This was quite plausible: the former Haitian police and army were dissolved after the UN intervention on the grounds of their abuse of human rights, and Wilfred's father had been a close political friend of the local magistrate. The magistrate told me that the uncle was beaten only after conviction, and indeed that the police had rescued him from an angry local mob.) The female cousin, who maintained good relations with the parents, denied any involvement in Wilfred's zombification but would not allow me to open the tomb: 'It's all finished now: justice has been done.'

I met Wilfred outside his parents' four-roomed house, the largest in the village and with concrete floors and galvanized iron roof. He was a slightly built man, constantly scowling, looking younger than his given age, much thinner than in an old photograph his parents showed me. In brief he was more or less mute, with limited spontaneous movement, rarely speaking and then only in single words. He could not describe his period of burial or enslavement – or indeed anything else – but agreed he was *malad* ('ill') and *zombi*. His parents reported that he would tell them when he was hungry, but they had to bathe him and change his clothes; he had no interest in anything. Wilfred's eyes continually scanned around him without clear intent, his fingers picking aimlessly at his nails or at the ground, and he avoided eye contact. His parents told me about periods of anger and irritation when he would ineffectually hit and kick out at others generally after being teased, and *malkadi* ('fits') during his sleep about once a week when he would cry out and his limbs would go in spasm. Wilfred did not talk to me spontaneously but would respond monosyllabically and appropriately to immediate practical questions; his mood was generally uninterested and irritable but he could be coaxed into acquiescence with sweets; he laughed when seeing me getting undressed for the night; some sullen suspicion of parents and of myself was evident.

On visiting his tomb with him, he did not display any evident emotion or desire to return to his grave, but reluctantly leant against it with a blank expression, becoming irritated when teased about his previous death by neighbours. My presumptive medical diagnosis was of an organic brain syndrome and epilepsy: his fits were reduced to once a month by giving him phenytoin. (The family had not initially taken him to a doctor because of the cost and because he was evidently a case of zombification.) I took blood from him and his family: subsequent DNA fingerprinting suggested that Wilfred was not the son of his putative parents. Nor was another zombi (whom I diagnosed with learning disability and foetal alcohol syndrome) related to her presumed family. As with Wilfred, Marie was afforded protection and care by this rescuing family, and in her case her presumed brother obtained wider support by converting to Pentecostalism and attempting to heal her. The case was reported on Radio Creole. In a third case, my colleague Chavannes Douyon made a diagnosis of catatonic schizophrenia.

Discussion

We cannot of course prove from a limited number of cases that in others an identified zombi might not have, as the Harvard Project suggested, a discrete pathophysiological state or at any rate one caused by some sort of poisoning.[10]

We are nevertheless left with two strong assumptions: (1) That a large number (or all) of recovered zombis are cases of mistaken identity, a point that has been suggested by earlier investigators (Mars 1945). Certainly the secret slave plantations of *zombis jardins* have never been identified, and we are left with only the odd locally recalled instance of a zombi in something resembling captivity (Davis 1988: 59, 62). (2) That locally identified zombis have a chronic mental illness, contrary to popular understanding. Zombis are characterized in Haiti by their extreme passivity, which is locally contrasted with the mentally ill: they cannot lift up their heads and have a nasal intonation (which they share with the Gèdè gods of death), a fixed staring expression, repeated purposeless and clumsy actions, and limited and repetitive speech. Unlike the mad they are easily subdued and controlled. In spite of the recognition of zombis in psychiatric hospitals and their apparent similarity to the mad, at least for the external observer ('a wretched lunatic', 'poor idiot girl' – Métraux 1958: 281; Huxley 1966: 36), they are supposedly distinguishable for the Haitian. Louis Mars (1945) too suggests recovered zombis are mentally ill, and Wade Davis (1988: 62) refers to another example. My two cases cited above both appear to have organic brain damage that, unlike psychosis, might tend to 'passivity'. In spite of the reported use of herbal medicine to treat madness (Huxley 1966: 48, 51, 122, 244, 246), the usual Caribbean understanding is that the condition involves spiritual matters – a spirit sent by a sorcerer, a failed conjuration, a spirit returned on to a sorcerer or an angry or jealous *lwa* often in the course of possession (Métraux 1958; Kiev 1961; Littlewood 1993). Treatment, generally ineffectual, would thus have to be spiritual counter sorcery or appeals to Le Grand Maitre. And with elementary mental health services, the destitute and wandering madman is, just like the *zombi savan*, a not uncommon sight along Haiti's roads and villages (Douyon 1972).

Zombis then appear to be socially rescued people who are mentally ill yet not recognized as such. Is the zombi complex to be understood as the unacknowledged recognition and restitution of mental illness? (And thus appear to be somewhat functional.) This seems unlikely for, as in Wilfred's case, the suspicion started at his death and not at his later retrieval: accusations seem more likely to occur immediately after sudden bereavement.[11] There is a common though fading Caribbean idea that no death is a 'natural' death (*mo bondye*) (e.g. Brodwin 1996). And healthcare systems are seen by us as a conscious response to a recognized problem (i.e. remove or confront the identified cause of the illness). The zombi only seems to be cared for because he is presumed to be a family member. And that the zombi is (to us) mentally ill seems irrelevant to local zombi care: the pattern presented just falls conveniently into what we might expect if one's *ti-bon anj* is removed – a loss of agency. And 'zombi care' seems quite nonspecific. It is rather then that the zombi complex laterally derives from other cultural needs: as the functionalists argued, it seems to explain the success of others and gives a voice to jealousy and suspicion. If the zombi complex only hints at maintaining social cohesion, then how are we to explain it? I should like to use an analogy here from evolutionary theory. There the 'motive power', so to speak, of the whole system is natural selection. Functionalism was to an extent a replaying of the theme of the earlier diachronic natural selection by the idea of synchronic social cohesion, both ways of explaining why some patterns persisted and some did not. If a pattern did not contribute to adaptive fitness – or in our case, social cohesion – it was eliminated in favour of variants through its own non-utility or actual harmfulness. Simply, its carriers would die out. But certain patterns could of course hang on if they were not drastically less successful either biologically or socially (and thence of course biologically – we do not find societies that practice ritual sacrifice of all the newborn). And here I want to evoke the analogy of the 'spandrel', as put forward by Steven Jay Gould, in evolutionary theory: in this, the spandrel is a characteristic that persists as an associated, but not in itself adaptive, by-product of something more fundamental (Gould and Lewontin 1979). This of course assumes that at the analytic level one can distinguish those patterns that are fundamental from those that are not and at any one time (for what is initially only a spandrel may by later changes in the environment become fundamental or modified into something more fundamental). And here we share with evolutionary theory the impossibility of carrying out experiments that would show how some elements are more essential while others are not.

The idea of the spandrel has already been used in social anthropological studies of the cognitive basis of religion to refer to something close to its evolutionary sense: religion as a social spandrel on the more fundamental process of, say, agent hyperidentification, which has immediate biological value. Here I am moving matters further into the social domain, so that the primary purpose of sorcery accusations (and hence zombification) is *social* cohesion (and thence the biological advantage is further removed) and the fact that zombis are mentally ill is just a spandrel on this. Clearly, the development of human culture includes spandrels on spandrels with declining social (or biological) adaptiveness – a further and further remove from any function. Ultimately, however, any social spandrel cannot be long inconsistent with social functioning. Why not let the zombi die (as with Australian sorcery victims – Eastwell 1982)? Whilst the social rescue of the mentally ill zombi seems decent and civilized from our own point of view, it hardly seems essential: there are too few zombis for this pattern to make much sense as part of a general social mechanism, nor are they offered any distinctive treatment. At the same time as our residual anthropological functionalism, we now however recognize – indeed, even privilege – social action: humans consciously make and remake their worlds, or at least, seek to do so. And this leads me briefly to ask a question much broader than the subject of this chapter: Why do societies have medical systems in which a great number of people (proportionately more than the population of zombis in Haiti) are cared for and optimistically treated? What are medical systems for?[12] And here we are likely to go for a more 'social action' type of interpretation – people are in pain and explicitly demand surcease. And that is generally distributed, but what of the restitution of those who cannot speak for themselves? Are we here with Marvin Harris's non-essential 'culture-by-whimsy' (1968: 567), or do we need to postulate some process or institution more closely related to a society's central functioning? A functionalist sort of model indeed. As Alfred Métraux (1958: 365) argued concerning *vodu*, 'In the anthropologist's sense, it functions.'

What does the zombi represent? Under the Duvaliers, who mobilized the *oungans* as their secret police (Diederich and Burt 1972; Nicholls 1996), and in the lengthy period of political terror, social instability and economic blockade during and after the Duvalier regime, with considerable individual mobility around a country with minimal internal communications except person-to-person contact (Thomson 1992, Aristide 1993), numerous case of abduction, torture, sexual slavery and secret homicide cloaked in *vodu* have still maintained an uncertain status between state terror and village-level suspicion of sorcery (Human Rights Watch 1996).[13] A more nuanced consideration of the significance of zombification requires a sociologically grounded analysis involving considerations of Haitian identity and of the wider political articulations of village-level sorcery accusation (cf. Taussig 1980). An as yet unexplored issue is how the agency-less zombi serves for the national history of Haiti, the 'Black Republic' of former slaves who have continued to face the ever-present threat of political dependency, external intervention and the loss of self-determination. Imagery and meanings of the Haitian Revolution are everpresent for contemporary citizens in art, media and public ceremonies.

Laennec Hurbon argues their *lwa* spirit is 'a foreign power, unknown and anonymous, that pounces on the individual who can only regain his or her balance by renewing a dialogue with him' (1995: 146). He uses the clinical data of Louis Mars to suggest that the Haitian madman is suspended between peasant traditionalism and the modern urban world.[14] The use of the whip in *vodu* to energize a charm (*wanga*), to order initiates and to dismiss gods, like the idiom of control over the hapless zombi, and the fact that zombis can be sold for plantation labour, all evoke Haitian slavery and its successors. The Haitian fear of zombification graphically represents the possibility of the loss of self-determination for the Haitian peasant. As anthropological interpretations, function and culture are hardly incompatible.

Roland Littlewood is professor of anthropology and psychiatry at University College London. He is a former president of the RAI and has undertaken fieldwork in Trinidad, Haiti, Lebanon, Italy and Albania, and has published eight books and around two hundred papers.

Acknowledgements

I would like to thank Professor Louis Mars for suggestions, Chris Ledger and Chantal Regnault for locating the zombis studied, 1996–97, and Dr Chavannes Douyon for helping me interview them. Chris Ledger played an essential role in facilitating my stay, and in suggesting and arranging the DNA fingerprinting.

Terminological Note

I use the word *zombie* (with an 'e') for the general Western concept, *zombi* (more properly *zonbi*) for the specifically Haitian phenomenon. And I translate *lwa* (African spirit or minor diety in *vodu*) as 'god'.

Notes

1. And gender, in its turn, is to be understood through medicine and warfare, etc. A similar schema is followed with the 'functionalism' of the cognitive philosopher, which argues that computations (on either electronic or neural substrates) are to be understood in terms of their causal relations rather than in their physical realization (Block 1980) – regarded metaphysically, mental states are merely functional states. Compare ontic structural realism in the philosophy of science (Ladyman and Ross 2008).

2. As Edmund Leach often remarked, we are all certainly functionalists in the field (Barnard 2000: 183).
3. Given our professional license to over-interpret (Olivier de Sardan 1999), there are few elements that cannot be imaginatively seen to fit, or become meaningful, in terms of other such elements. Only academic convention tells us when we are getting ridiculous (Littlewood 2007).
4. Her collection of papers placed together as 'witchcraft' both the intrinsic capacity to harm others, involuntary, and the technical use of substances or spells to do the same – respectively, 'witchcraft' and 'sorcery' in Evan-Pritchard's Zande translations. The volume thus included Ardener's paper on Cameroon *nyongo* sorcery, akin to zombification in Haiti. Classical African witchcraft does occur in the Caribbean in the form of night-flying witches within the community who take the form of animals to suck the blood of their victims – as *soucouyants* or *lagahoos* in Trinidad, *loups garous* in Haiti and Guadeloupe.
5. As in MacFarlane's 'charity denied' model of seventeenth-century English witchcraft.
6. The Americans did establish agriculturally orientated schools, but the mass of the peasants were compelled to provide government work on the roads or plantation labour. One local contemporary terms the reintroduction of the plantation system as 'the legalised assassination of the Haitian rural population', with the production of a wandering landless proletariat as 'economic slavery' (Nicholls 1996: 150).
7. Louis Mars et al., personal communications. My original co-author, Chavannes Douyon, told me that his brother Lamarque, who was one of the collaborators on the Harvard Project (and who had trained in psychiatry at McGill in the 1960s during its period of psychotomimetic enthusiasm), subsequently died and was locally assumed to have been turned into a zombi.
8. We are unlikely to find coherence and consistency in the local schemata. As Mauss (1950: 88) observes: 'The normal condition of magic is one involving an almost total confusion of power and order.' Different informants in parts of Haiti maintain rather different ideas. Indeed Métraux, working in Marbial not far from the zombis I describe below, says that the very distinction between *gwo-bon anj* and *ti-bon anj* 'is often forgotten' (1958: 257). Nevertheless, he often talks of the former as something recalling our idea of an earthbound ghost, the latter something more of an ethereal soul. His *gwo-bon anj* wanders in sleep, is displaced by a *lwa* (vodu god) during possession or in madness, or after accidents, and during severe trauma; if of an initiate it can be stored safely by the priest in a jar; after death it is retrieved and similarly put in a jar on the altar of the vodu temple where it can become a minor *lwa*. So the apparent distinction between the *gwo-bon anj* as a sort of vital animal soul with the *ti-bon anj* as something recalling a Christian soul does not fully hold. To add to the confusion, a protective *lwa* can be 'fixed' in the initiate's head (*lwa-met tet*) without apparently a loss of either *bon anj*; if that is done, the possessing god has to be removed after death. Nor can we establish a consistent link or distinction between death, zombification, possession and madness; there are some common features of all four, both in operation and appearance – a spirit for instance can send you mad and a disembodied *bon anj* used after death for sorcery may be referred to as a zombi. The essential point is that zombi and madman can be locally distinguished on meeting with them, and that zombis are characterized by their loss of agency (compare Métraux 1958: 99, 131; Huxley 1966: 103; Davis 1988: passim). Local Haitian doctors suggested to me that zombification can be recognized only by the absence of any characteristic features of mental illness, yet with reduced awareness and verbal and motor perseveration.
9. But see Métraux 1958 for accounts of dangerous zombis: these are presumably captured *zombis astrals* that can be used for spirit attack by sorcerers. But Davis (1988: 61) says the *zombi cadavre* when fed salt can turn on his master.

10. Given that death is locally recognized without access to medical certification, and that burial usually occurs within a day of death, it is not implausible, if presumably rare, for a retrieved person to be still alive. In some cases they might be well enough to recover with minimal care. The use of datura to 'revive' them, and its possible repeated administration during the period of zombi slavery, could produce a schizophreniform state of extreme passivity. The number of *bòkòs* who told me they were engaged in attempts at zombification may suggest the breaking open of tombs by an optimistic sorcerer in the belief that this has been achieved. The use of human remains in sorcery is so common that many country tombs have been broken into, and the majority of *oufos* (temples) I visited contained retrieved human skulls and other body parts.
11. Compare *mu-ghayeb* in Oman where an assumption that the dead person is only ensorcered and will return alive seems to help deny the loss altogether (Al-Adawi, Burjorjee and Al-Issa 1997).
12. Malinowski argued that health was a 'basic need' that called forth the social response of 'hygiene' (Kaberry 1957) – which does not really get us very far. Its simplistic functionalism seems to lead to what Douglas ironically termed the identification of 'Moses as a public health inspector'. By contrast, Fabrega (1997) argues that medical systems are elaborations of a 'sickness/healing adaptation' that is biologically efficacious and whose antecedents in our hominid ancestors may be inferred from certain pongid analogues – the succouring of conspecifics, the recognition of 'time off work', the use of various plants to self-medicate and treat others, removal of maggots from festering wounds, and the use of leaves to dab and clean them – a pattern of reciprocal altruism that becomes part of developing human sociality and both increases the number of productive individuals and helps avoid group frictions consequent on crises of sickness and accidents. Interestingly Fabrega (1997: 43–44) suggests that this became established along with the possibilities of feigning illness (malingering).
13. Dr François Duvalier was of course an early medical anthropologist (e.g. Denis and Duvalier 1944), and presumably the first to become a god (*lwa*).
14. Mars, the doyen of Haitian psychiatry (Farmer 1992b), told me the idea is not completely fanciful. That the madman of the English-speaking Caribbean serves as a local image of the hapless postcolonial subject has been suggested by both Fisher (1985) and Littlewood (1993). If we pursue this line, then the zombi is simply an extreme, bracketed-off, instance. And in that, we may discern a sort of social functionalism (group cohesion).

References

Al-Adawi, S., R. Bmjoijee and I. Al-Issa. 1997. 'Mu-ghayeb: a Culture-specific Response to Bereavement in Omar.' *International Journal of Social Psychiatry* 43: 144–51.

Anon. 1984. 'Puffers, Gourmands and Zombification' (editorial). *Lancet* 323: 1220–21.

Ardener, E. 1970. 'Witchcraft, Economics and the Continuity of Belief.' In M. Douglas (ed.), *Witchcraft Confessions and Accusations*. London: Tavistock, pp. 141–60.

Aristide, J. B. 1993. *Tout Homme est Un Homme, Tout Moun se Moun*. Paris: Seuil.

Barnard, A. 2000. *History and Theory in Anthropology*. Cambridge: Cambridge University Press.

Block, N. 1980. 'Introduction: What is Functionalism?' In N. Block (ed.), *Readings in the Philosophy of Psychology*. Cambridge, MA: Harvard University Press, pp. 171–84.
Brodwin, P. 1996. *Medicine and Morality in Haiti: The Contest for Healing Power.* Cambridge: Cambridge University Press.
Davis, W. 1988. *Passage of Darkness: The Ethnobiology of the Haitian Zombie.* Chapel Hill: University of North Carolina Press.
Denis, L., and F. Duvalier. 1944. 'L'Evolution Stadiale de Vodou.' *Bulletin du Bureau d'Ethnologie* 3: 9–32.
Diederich, B., and A. Burt. 1972. *Papa Doc: Haiti and its Dictator.* Harmondsworth: Penguin.
Douglas, M. 1970. 'Introduction.' In Douglas (ed.), *Witchcraft Confessions and Accusations*. London: Tavistock, pp. xiii–xxxviii.
Douyon, L. 1972. 'Introduction aux Traitements des Malades Mentaux en Haiti.' *Bull Centre Psychiatric Neurology* II: 5–8.
———. 1980. 'Les Zombies dans le Contexte Vodu et Haitien.' *Haiti Santi* 2: 19–23.
Eastwell, H. D. 1982. 'Voodoo Death and the Mechanism for the Dispatch of the Dying in East Arnhem.' *American Anthropologist* 84: 5–18.
Evans-Pritchard, E. E. 1937. *Witchcraft, Oracles and Magic Among the Azande.* Oxford: Clarendon Press.
Farmer, P. 1992a. *Aids and Accusation: Haiti and the Geography of Blame.* Berkeley: University of California Press.
———. 1992b. 'The Birth of the Klink: A Cultural History of Haitian Professional Psychiatry.' In A. D. Gaines (ed.), *Ethnopsychiatry*. New York: State University of New York Press, pp. 251–72.
Fabrega, H. 1997. *Evolution of Sickness and Healing.* Berkeley: University of California Press.
Fisher, L. E. 1985. *Colonial Madness: Mental Illness in the Barbadian Social Order.* New Brunswick, NJ: Rutgers University Press.
Geschiere, P. 1998. 'Globalization and the Power of Indeterminate Meanings: Witchcraft and Spirit Cults in Africa and East Asia.' *Development and Change* 29: 811–37.
Gould, S. J., and R. C. Lewontin. 1979. 'The Spandrels of San Marco and the Panglossian paradigm: a Critique of the Adaptationalist Programme.' *Proceedings of the Royal Society of London* 205: 561–98.
Harris, M. 1968. *The Rise of Anthropological Theory.* London: Routledge and Kegan Paul.
Human Rights Watch 1996. *Thirst for Justice: A Decade of Impunity in Haiti.* New York: HRW.
Hurbon, L. 1993. *Vodu.* Paris: Gallimard. English edition (1995), London: Thames and Hudson.
Hurston, Z. N. 1938. *Voodoo Gods.* London: Dent.
Huxley, F. 1966. *The Invisibles.* London: Hart-Davis.
Kaberry, P. 1957. 'Malinowski's Contribution to Field-work Methods and the Writing of Ethnography.' In R. Firth (ed.), *Man and Culture: An Evaluation of the Work of Bronislow Malinowski*. London: Routledge and Kegan Paul, pp. 71–92.

Kiev, A. 1961. 'Folk Psychiatry in Haiti.' *Journal of Nervous and Mental Disease* 132: 260–65.
Ladyman, J., and D. Ross. 2008. *Everything Must Go: Metaphysics Naturalised.* Oxford: Oxford University Press.
Littlewood, R. 1993. *Pathology and Identity: The Work of Mother Earth in Trinidad.* Cambridge: Cambridge University Press.
——— (ed.). 2007. *On Knowing and Not Knowing in the Anthropology of Medicine.* Walnut Creek, CA: Left Coast Press.
Littlewood, R., and L. Douyon. 1997. 'Clinical Findings in Three Cases of Zombification.' *Lancet* 350: 1094–96.
Mars, L. 1945. 'The Study of Zombie in Haiti.' *Man* 45: 38–40.
Mauss, M. 1950. *A General Theory of Magic.* English edition (1972). London: Routledge and Kegan Paul.
Métraux, A. 1958. *Le Vaudou Haitien.* Paris: Gallimard. English edition (1959), London: Deutsch.
Nicholls, D. 1996. *From Dessalines to Duvalier.* 3rd edition. London: Macmillan.
Olivier de Sardan, J. P. 1999. 'Interpréter, surinterpréter.' *Enquete* 3: 22–43.
Radcliffe-Brown, A. R. 1977 [1933, 1937]. '"Function", "Meaning" and "Functional Consistency".' In A. Kuper (ed.), *The Social Anthropology of Radcliffe-Brown.* London: Routledge and Kegan Paul, pp. 43–52.
Rivers, W. H. R. 1914. *The History of Melanesian Society: I.* Cambridge: Cambridge University Press.
St. John, S. 1884. *Hayti or the Black Republic.* London: Smith, Elder and Co.
Stocking, G. W. 1984. *Functionalism Historicized: Essays on British Social Anthropology.* Madison: University of Wisconsin Press.
Taussig, M. T 1980. *The Devil and Commodity Fetishism in South America.* Chapel Hill: University of North Carolina Press.
Thomson, I. 1992. *Bonjour Blanc.* London: Hutchinson.

Chapter 10

Religion and Psychosis

A Common Evolutionary Trajectory?

Simon Dein

> O Lord, you have searched me and you know me. You know when I sit and when I rise; you perceive my thoughts from afar, you discern my going out and my lying down, you are familiar with all my ways.
> – Psalm 139:1–3 NIV

Explanations deriving from evolutional psychology have often attracted criticism for post hoc storytelling and the fact that evolutionary hypotheses are unfalsifiable. Nevertheless, I would argue that evolutionary psychology plays a central role in transforming psychology from a largely atheoretical collection of findings to a discipline that accounts for why the components of the mind/brain have the designs they do. Evolutionary explanations have been invoked to make predictive hypotheses that have been subsequently empirically confirmed (Sell et al. 2003). Included are such areas as social exchange and cheater detection (Verplaetse, Vanneste and Braeckman 2007), foraging and sex differences in spatial ability (Silverman and Eals 1992), race encoding as a byproduct of coalition encoding (Kurzban, Tooby and Cosmides 2001) and postpartum depression as an adaptation to reduce investment in the newborn in the context of limited resources (Tracy 2005). Here we focus on the rapidly growing area of the evolution of religion and its relevance for the contemporary diagnostic category of schizophrenia. The contribution is to suggest similarities in the cognitive foundations of both religion and schizophrenia.

Notes for this chapter begin on page 206.

That religious experiences and psychosis (more particularly schizophrenia) share similar, or indeed identical, psychological characteristics has often been maintained since suggestions in classical Greece (Simon 1978). In both popular and medical perceptions, religious enthusiasm and experience have often been equated with madness. From the eighteenth century onwards, scholars have often argued that the founders of new religious dispensations might be conspicuously unusual or even frankly insane (Littlewood 1993).

The debate on the mental health of the shaman in particular has long continued among social anthropologists, some of whom have even ventured that all religious cognition is psychotic (e.g. La Barre 1970). In terms of empirical study, the evidence for this hypothesis has of course been problematic, given the difficulty of providing a conventional psychiatric assessment of the religious leader as they develop a new (or modified) religious dispensation. An exception is the founder of a new religion in the Caribbean, studied by Littlewood (1993). It has generally proved easier, if less reliable, to offer a retrospective assessment of religious leadership based upon existing biographical sources (e.g. Littlewood 1996). Anthropological accounts of religious experience and certain aspects of psychopathology in contemporary Euro-Americans have argued that, from the phenomenological perspective, they may be identical (Jackson and Fulford 1997; Dein and Littlewood 2007). Any differentiation depends on the social consequences: in William James's reframing of Jesus's 'By their fruits you shall know them' (1902: 36).

To develop an argument on the convergence of religion and psychosis, we need to: (1) use coherent definitions of both, and to restrict 'religion' here to some Neo-Tylorean mentation ignoring for the moment the social institutions and doctrines which insubstantiate this; (2) argue that they are essentially the same process, or that one is primary and the other is inherently piggy-backed onto it, or that both follow closely on some other third phenomenon in a way that other patterns, social or psychological, do not; and (3) if we are arguing a long term convergence, then we can offer appropriate evolutionary explanations, either social or biological selection (or both).[1]

Several suggestions have been offered to explain why schizophrenia (taken as cross-culturally found) persists and why, with its lower fertility, it has not disappeared through natural selection. One possibility is that schizophrenia has an evolutionary advantage either for the individual's immediate kin or the larger group in terms of reproductive success or survival: Erlenmeyer-Kimling and Paladowski (1968) have found that female infants of parents with schizophrenia surprisingly enjoy increased survival compared to other children. J. L. Karlsson (1984) argues that the evolutionary advantage of schizophrenia in Western societies lies in the enhanced creativity (and hence survival) of relatives. T. M. Crow (2000), arguing that the evolution of schizophrenia and language are intimately related, suggests a possible lack of cerebral asymmetry observed in schizophrenia along with its characteristic alterations in language use: as language per se confers selective advantage, then schizophrenia is dragged along too.

Anthony Stevens and John Price (2000) propose a functionalist model of religion in which an expanding small-scale ancestral community must eventually split to conserve a population size optimal for its ecological niche: charismatic leaders with schizotypal traits use 'paranoia', 'delusions', 'religious scenes' and neologisms to control and fractionate groups by seeking new social dispensations and settings. For Joseph Polimeni and Jeffery Reiss (2002), schizophrenia, or something allied to it, could enhance the leader's ability to initiate and conduct religiously based rituals: such rituals are universally observed in all cultures and thus are likely both to be genetically rooted and perhaps critical for survival, presumably in terms of group cohesion and mobilization.[2] They argue that until the past few thousand years humans have always lived in hunter-gatherer societies with some form of 'shamanic' leadership. Psychosis might be advantageous for these individuals in generating novel and persisting religious rituals from an 'altered state of consciousness'. Due to a lack of contemporary empirical evidence and their extremely conjectural speculation, both these theories, like other social evolutionist scenarios, must remain extremely problematic.

One may ask whether what one terms 'religion' is an adequate general category for comparison. Religion is generally regarded as a shared cultural institution whereas schizophrenia is fundamentally an individual experience. Religious experience may take numerous forms, including feelings that all things are one, a sense of transcending time and space, a feeling of the holy, sacred and divine, and a sense that such events and situations cannot be described in words (Beit-Hallahmi and Argyle 1997). Here we adopt the view that religion primarily involves recognition of the agency of ultrahuman agents, agreeing with Pascal Boyer (1994: 9) that the notion of ultrahuman entities who posses something like human agency seems to be the only evident universal found in religious cosmologies. Actual religious practice of the sort studied by social anthropologists is always organized around such agents and it is upon these that seem to be erected systematized reflection (doctrine), prescribed behaviours (rituals and so on) and communal structure (social organization). The acceptance of such agents may be the most commonly offered definition of religion in the social sciences (Sosis and Alcora 2003). For religious adherents themselves, gods do not just exist, they *matter* to those who believe in them and they render everyday events significant through their association with a divine cosmology. I propose, along with Boyer (2001), that religious cognition is a specific form of cognition characterized by a focus on ultrahuman agents; it is counterintuitive and it is costly in terms of time and emotional involvement.

That this sort of cognition might be related to schizophrenia is suggested by various findings. First, in schizophrenia, there is a substantial occurrence of religiously oriented delusions in all societies examined (Brewerton 1994; Moslowski et al. 1998; Siddle et al. 2002). Second, religious ideas and assertions of the 'paranormal' are especially common among those individuals in the West with schizotypal traits (Thalbourne and Delin 1994). Third, there are phenomenological parallels between schizophrenic and normative religious hallucinations (Dein and Littlewood 2007). Finally, there is emerging evidence for a continuum between religious normality and psychosis: members of new Euro-American religious movements have been found to have similar scores on various 'delusion scales' as those who have been diagnosed clinically with psychotic illnesses (Peters et al. 1999).

Shared Modules?

Can such a postulated link between 'religion' and 'schizophrenia' be regarded as intrinsic? There seems some evidence that religious thinking in schizophrenia is a strategy to cope with extreme experience (Mohr et al. 2006). The usual explanation is that religious beliefs and practice may provide a source of fundamental meaning, solace and stable purpose in a radically confusing world. Here we propose a rather different theory in which 'religion' and schizophrenia both represent applications or by-products of certain mental modules, which once had some adaptive advantage. Such modules include: human agency, our sense of personally owning, controlling or identifying with an experience or action, and what psychologists term 'theory of mind' (ToM), our ability to attribute mental states such as cognition, intention and agency to other people, thus allowing the social individual to explain, predict and manipulate the behaviour of others. The two modules are related, for other recognized agents are presumed to act via intentions, motivations and desires just like ourselves. Both religious cognition and schizophrenia, compared with everyday cognitions, involve agency overdetection and overextended ToM. In everyday religious cognition, agency detection and ToM modules function fairly normally; in schizophrenia, as we shall discuss below, both modules are impaired.

While religious cognition is based upon an intact ToM and agency-detection mechanism, under certain circumstances (e.g. rituals, dancing, music) the ToM functions in a way that results in a breakdown of the boundary between the self and outside world. Schizophrenia and 'extreme' religious experiences[3] both involve breaches in our (Western) boundaries between self and 'other'. In both there is a subjective breach of a perceived psychological border to the self, conventionally termed the 'ego-boundary', a breach that in psychosis is experienced fairly concretely (Jaspers 1926). In some way the subjective self becomes more permeable, and patterns of influence and agency may pass 'into' or 'out from' the self depending on the symptom type: '[The] patient knows that his thoughts and actions have an excessive effect on the world around him, and he experiences activity, which is not directly related to him, having a definite effect on him' (Fish 1967). In a similar way to this 'magical thinking' (as it was once called), in many religious experiences the boundary between self and divine is breached. This occurs in the extreme during mystical states where the self seems to be completely absorbed into the divine.

Although we suggest a cognitive mechanism shared between religion and schizophrenia, we are not arguing for a shared identity or that religion *is* a form of mass delusion as has been suggested by some (La Barre 1970; Schumaker 1995; Dawkins 2008).

Agency and Religion

Religion in an extended sense is virtually universal and hence perhaps 'natural' (Boyer 1994). Following Stephen Jay Gould (1997), the evolutionary biologist who popularized the notion of a spandrel (an associated by-product of evolution itself not immediately adaptive), Boyer (1994), Rob Barrett (2000) and Scott Atran (2002) have offered a set of arguments, now termed the 'cognitive science of religion', that attempts to describe the primary mental procedures upon which religious ideas are 'parasitic'. Religious cognition is seen as dependent upon psychological faculties that have evolved for other purposes, and is not immediately adaptive in itself. Religion is thus an extraordinary use of everyday cognitive processes that cannot exist apart from the individual minds and in environments that can employ them.

What gets newly transmitted in social life must fit well with existing patterns of cognition. Religious representations, especially those of deities with something like a mind, seem easily represented by the brain, for they rest upon the cognitive processes that have developed for more immediate advantages. They have properties that fit with existing natural ontological categories such as personhood. They do differ from these categories in some respects but only slightly, and are thus described by Boyer (1994) as minimally counterintuitive. For instance the gods have wishes and emotions just like humans but they also possess novel qualities extended from everyday life like omnipotence and omniscience – or, in the case of ancient Egyptian religion, a falcon's head. Shades have a form yet they pass through walls. It is these counterintuitive properties, argues Boyer, that renders representations salient and hence memorable and persisting.

Agent Detection and Religious Representations

Atran (2002) is typical of this position: he regards religion as a spandrel on the cognitive mechanism responsible for agent detection, which, he argues, developed in evolution for inferring the presence of dangerous predators. A large psychological literature suggests that to attribute agency at times to nonhumans and to inanimate objects is universal (Heider and Simmel 1944; Michotte 1963). Agency overgeneralization may be an innate feature of human cognition; even infants may see inanimate movement as purposive behaviour (Gergely and Csibra 2003). According to Atran our brains are primed to identify the presence of agents, even when such presence is unlikely. Barrett (2000), like Boyer (2001), argues that we have a hyperactive agent detection template that seeks out the presence and intentions of other beings around us. Boyer argues this is represented in children's imaginary companions, which provide representations for the social mind. Similarly, Atran and Ara Norenzayn argue that 'widespread counterfactual and counter intuitive belief in supernatural agents could be explained by the fact that they trigger a naturally selective agency detection system, which is trip wired to respond to fragmentary information, inciting perception of figures, lurking shadows and emotions of dread and awe' (2004: 714). A by-product then is the susceptibility to infer ultrahuman beings. Ultranatural agents are readily conjured up because natural selection has resulted in a cognitive mechanism for privileging agency detection in the face of uncertainty: it is clearly more adaptive to mistakenly identify a shadow as a predator than the reverse. And the identified agency of another will be elaborated as akin to our own experience of personal agency: in other words we take our own experiences and perceive them in others ('projection').

These ideas accord with Stewart Guthrie's (1993) ideas of the origins of 'animism' in anthropomorphism,[4] which he then uses as an explanation for established religion. He too postulates that in the 'environment of evolutionary adaptation' it would have been beneficial for humans to be able to quickly and easily signal out the presence of other people and animals that might present danger. His examples of the everyday overextension of this include hearing voices in the wind or seeing faces in the clouds (ibid.).

If postulation of the existence of supernatural beings might relate to the misattribution of agency at certain times, it does not immediately explain the *persistence* of religion. Once suggested and organized, the gods then have social advantages for us in maintaining group cohesion and mobilization as the ultimate underpinnings of a culture in terms of norms and sanctions, and indeed in the very possibility of meaning open to a society.

Theory of Mind, Metarepresentation and Religion

Theory of mind allows us to anticipate the actions of others by imagining ourselves in another's place: what would X do if X was like me? Boyer (2001) argues that by developing a hypertrophied theory of mind we are able to predict the behaviour of others and thus facilitate cooperation and social interaction. Sociality is dependent upon predicting and anticipating the behaviour of others with a fair degree of accuracy. The essence of ToM, its central adaptive importance and the reason why it has evolved, is that this adaptation is primarily concerned with making inferences concerning the dispositions, motivations and intentions of others.

Ultrahuman agents too are intimately involved in interactions with people: as in human interactions, they employ the processes involved in social cognition. Supernatural agents do not just act, they are represented as having *knowledge* of humankind and as thus themselves possessing a theory of mind in attributing mental states such as belief and intention to people. As Bloom (2004) has argued, if one accepts the idea of minds (which cannot be empirically verified), it is a short step to postulate minds that do not have to be anchored in bodies, and thence another short step to both an immaterial soul and a transcendent deity.

Gods and spirits are not represented as having human features in general, but as having minds: they are held to have thoughts, memories and emotions and perceive events. Although corporeal features, such as having bodies, eating food or aging, may also be attributed to them, the only feature of humans that is invariably projected onto supernatural beings is the mind (Boyer 1994).

Ultrahuman agents do not merely possess minds, they deploy these minds in interacting with human agents. Boyer argues that the special properties of supernatural agents that mean they matter to people are those that directly activate 'mental systems geared to describing and managing social interaction with other human agents' (2002: 77). In normal social interactions, people make moves, the consequences of which then depend on the moves of others. Unlike humans who have limited access to the intentions of others, supernatural agents have full access to strategic information – the various sources of input that the social mind uses to evaluate a particular individual or situation and to influence ongoing social interaction.

Todd Tremlin argues 'in every culture the gods that matter know the truth, keep watch, witness what is done in private, divine the causes of events and see inside people's minds' (2006: 115). In Judaeo-Christian sacred texts several passages attest to the fact that God can read our minds. Psalm 139:4 (NIV) argues, 'Before a word is on my tongue you know it completely, O Lord.' Jesus as the incarnation of a supernatural agent is held to know people's thoughts (Mark 2:6–8). According to Ezekiel (11:5, NIV), 'Then the Spirit of the Lord came upon me, and he told me to say: "This is what the Lord says: That is what you are saying, O house of Israel, but I know what is going through your mind."' St Mark (13:11, NIV) describes the reverse, an external control over our thought and speech: 'Whenever you are arrested and brought to trial, do not worry beforehand about what to say. Just say whatever is given you at the time, for it is not you speaking, but the Holy Spirit.' Raffaele Pettazzoni argues for such divine omniscience being ubiquitous in religious systems:

> Divine omniscience has another field of activity; besides the deeds and besides the words of mankind, it examines even their inmost thoughts and secret intents. In the prophecies of Jeremiah we are told that the Lord tries 'thereins and the heart' (Jer. XI, 20). The same thought is found among many other peoples, savage and civilized. Karai Kasang, the Kachin Supreme Being, 'sees' even what men think. The Haida say that everything we think is known to Sins Sganagwa. The Great Manitu of the Ankara knows everything, including the most secret thoughts. Tezcatlipoca knows men's hearts; Temaukel, the Supreme Being of the Ona-Selknam, knows even our thoughts and most private intentions. In Babylonia, the god Enlil knows the hearts of gods and men, and Shamash sees to the bottom of the human heart. Zeus likewise knows every man's thought and soul. (1955: 20)

The idea that ToM may be a prerequisite for religious belief has been forcefully argued by Robin Dunbar (2005). The notion of a God who is just, who watches over us, who punishes us but who can admit us to Paradise depends on the understanding that other beings – in this case a supernatural one – can have human-like thoughts and emotions. Dunbar proposes that several levels of metarepresentation may be required, since religion is a social activity, which depends on shared beliefs.

The Earth People of Trinidad revere a notion of the divine as sensate Nature, but this is a Nature who has a memory of Her own and of human experiences, who has suffered and who is seeking, through cataclysm, to return the whole universe into a more just and orderly state (Littlewood 1993). As divinity, She knows what humans are thinking and planning, and will respond to them: She is now impatient. (Whilst its immediate origin was in a personal psychotic identification, the idea of a divine Nature here recalls a Caribbean Christianity creolized with contemporary ecological ideas.)

We would argue that the application of ToM is not limited to living beings whether real or imaginary, but rather it can extend thus to the natural world as well. Although agency may at times be attributed to inanimate objects, this is rare in the universalist religions. But the universe is imbued with purpose and intention. Are we cognitively predisposed, like the Earth People, to view the physical world teleologically? As Boyer (2001) puts it, 'Physical events around us are not just one damn thing after another; there often appear to be causes and effects. But you cannot claim a cause, at least literally. What you see are events and your brain interprets their succession as cause plus effect.' As Hume argued, we cannot see cause or purpose, all we actually see is 'one damn thing after another', but our brain takes these observations and infers that the things and events around us are purposeful. Religious people are predisposed to infer purpose when observing the natural world and this same process renders events as intended and as personally significant.

Among the Earth People, as among others, events like thunderstorms or the passage of strangers are immediately given meaning as divine incidents that are messages to the group of ultimate significance: always dependent, observes the ethnographer, on certain current debates, preoccupations and tensions among the group themselves (Littlewood 1993).

Alienation of Personal Agency in Religion and Schizophrenia

This distinction between self and other has been emphasized by the modern Western phenomenologists of religion. Rudolf Otto (1958) describes the 'numinous' in religious experience as the ineffable core of religion: a dependence on something greater than ourselves. Durkheim describes 'the man who lives according to religion . . . as above all a man who feels within himself a power of which he is not usually conscious, a power which is absent when he is not in a religious state' (Pickering 1994: 192). Similarly in the nineteenth century, Ludwig Feuerbach (1967) had talked of this self/other dimension rather more psychologically when he claimed, 'the ultimate secret of religion is the relationship between the conscious and unconscious, the voluntary and involuntary. The I and the not I in one and the same individual.' In a similar way James proposed surrender of the self as the essence of religion (1902: 403–4): 'At the same time the theologian's contention that the religious man is moved by an external power is vindicated, for it is one of those peculiarities of invasion from the sub-conscious region to take on objective appearances, and to suggest to the subject an external control.'

James (ibid.) describes here the loss of personal agency during mystical experiences: something he refers to as 'passivity' (as do contemporary psychiatrists). The oncoming of mystical states may be facilitated by preliminary voluntary operations, as by fixing the attention, or going through certain bodily performances, or in other ways that manuals of mysticism prescribe; yet when the characteristic sort of consciousness has once set in, the mystic feels as if his own will were in abeyance, and indeed sometimes as if he were grasped and held by (or absorbed into) a superior power. While few religious experiences attain the degree of unification associated with the mystical state, we would contend that all religious experience involves an alteration in the sense of agency to varying degrees.

The founder and leader of the Earth People is understood as a partial incarnation of the divine Nature in one human: like (the plausible example of) Jesus, she alternates between her mundane being with habitual everyday memories and actions, and the identification with supreme divinity – omniscient, powerful and frequently irascible (Littlewood 1993).

Beyond recent interest in their putative biological correlatives (Bartocci and Dein 2005), there has been little empirical attempt to look at the mechanisms involved in such religious experience. Tanya Luhrmann (2005) appeals to a psychological phenomenon of 'absorption' to account for both spiritual experience and dissociative disorders: related both to dissociation and hypnotic states, this is the capacity to become absorbed in inner sensory stimuli and to lose a degree of external awareness. There is evidence from ethnography (anthropological accounts of spirit possession), psychology (experimental work on hypnosis and meditation) and religious history (accounts and manuals of Christian mysticism) that the practice of 'absorption' can go along with the development of hallucinations; as the state of absorption deepens there is also a shift in the sense of agency so that individual mental events come to happen *to* a person from the outside (similar to the earlier views of James and G. H. Mead). Or indeed, at a less intense level, the indigenous psychologies and daily life of communities such as the 1940s Dinka (Lienhardt 1962).

Hyper-vigilance to perceived threat has been frequently cited as an explanation of paranoia. Individuals with paranoid delusions are especially sensitive to possible threats. Compared to non-deluded psychiatric patients or other people, those with persecutory delusions preferentially attend to threat-related stimuli and preferentially recall threatening episodes, and attribute failures to the malevolent actions of others (Ullman and Krasner 1969; Bentall and Kaney 1989).

There is some evidence that schizophrenia is a phenomenon primarily of the misattribution of agency. Patients with schizophrenia often report the immediate experience of someone else controlling their thoughts and actions. Alternatively they may feel that they are in control of external events or are convinced they know what other people think. Things or events are related to them in a special way or have a personal significance. Symptoms that suggest misattribution of agency include thought insertion and withdrawal experiences, passivity and certain auditory hallucinations (Fish 1967). In thought insertion the patient experiences thoughts that do not have the quality of being his own. In thought withdrawal the patient describes his thoughts being taken from his mind by an external force. Thought broadcasting is the experience that an individual's thoughts are not contained within his own mind: the thoughts escape from the confines of the self and may be experienced by those around. This may form the basis of the delusion that 'thoughts are being read'.

Among the wide range of manifestations in schizophrenia, these so-called first-rank symptoms are traditionally considered critical for its diagnosis. They are often seen to be primary. Kurt Schneider (1955) argues that these refer to a situation where patients interpret their own thoughts or actions as made by alien forces or other people. These symptoms reflect disruption of a mechanism that normally generates consciousness of one's actions and thoughts and allows direct attribution to their author alone. In schizophrenia patients may attribute to others rather than to themselves their own actions or thoughts; or in contrast attribute to themselves the actions or thoughts of others. Pierre Janet (1937) argued that the false attribution reflected personal representations of others' actions and thoughts in addition to the usual representation of one's own thoughts and actions. False attributions were due to the imbalance of the two representations. Those experiencing hallucinations thus misattribute their own intentions or actions to external agents.

Chris Frith (1992) suggests that an internal monitoring deficit causes delusions of alien control: these abnormal experiences arise through a lack of awareness of intended actions. Such impairment might cause thoughts or actions to become isolated from the sense of will normally associated with them. Internally generated voices or thoughts might be interpreted as external voices (auditory hallucinations and thought insertion) and ones actions and speech might be construed as externally caused (passivity or delusion of control).

The presumed phenomenological parallels between religious experience and schizophrenia may be particularly apparent in Western cultures. Horacio Fabrega (1982: 56–57) discusses how Schneider's first-rank symptoms depend on a 'Western' notion of self:

> These [first-rank] symptoms imply to a large extent persons are independent beings whose bodies and minds are separated from each other and function autonomously. In particular, they imply that under ordinary conditions external influences do not operate on and influence an individual: that thoughts, are recurring inner happenings that the self 'has'; that thoughts, feelings, and actions are separable sorts of things which together account for self identity; that thoughts and feelings are silent and exquisitely private; that one's body is independent of what one feels or thinks; and finally that one's body, feelings and impulses have a purely naturalistic basis and cannot be modified by outside 'supernatural' agents . . . and it is based on this psychology (i.e. a Western cultural perspective) that schizophrenic symptoms have been articulated.

Rob Barrett (2004) emphasizes the fact that a comparative cultural phenomenology requires ethnographic underpinnings. He describes the difficulties of studying these first-rank symptoms cross-culturally. Using data from the Iban of Borneo, he suggests that questions concerning auditory hallucinations translate with ease from English to Iban. In contrast, problems with thinking (thought withdrawal, insertion and broadcasting) make little sense in this group: unlike Western notions of thought as 'interior' and 'privatized', among the Iban thinking is partly a bodily process and partly an interactional process. A cultural phenomenology of 'unusual' experiences must take account of indigenous conceptualizations of mind and self.

Within Western cultures there are a number of counter suggestions to the ideals of mental autonomy and privacy. Such instances include telepathy and hypnosis, but more pervasive is the Christian belief in an omniscient God who knows inner thoughts. In the Christian tradition individuals can be punished or feel guilty or ashamed of these thoughts even if they are not verbalized. The experience of God's omniscience might act as a cultural model for articulating disorders of the privacy of thinking. Theology and the phenomenology of the self are dialectically related. We might speculate that the Reformation with its interiorization and privatization of religion might have facilitated the emergence of a bounded self and subsequently set the cultural grounds for the experience of schizophrenic thought disorder as we know it, especially those delusions and hallucinations that relate to loss of self-other boundaries.

Theory of Mind and Metarepresentation in Schizophrenia

Patients with schizophrenia have specific difficulties in inferring what others intend, think or pretend, and this ToM impairment probably influences the way they use language and interpret speech (Brune 2003). For instance, delusions of alien control and persecution, the presence of thought and language disorganization and other behavioural symptoms may be understood in light of a disturbed prior capacity to relate personal intentions to executing behaviour and to monitor others' intentions. A common symptom in schizophrenia is the delusion of reference: the phenomenon that external events are experienced as personally meaningful and specifically related to the patient. Mundane events may be held to be personally significant and to be 'set up' specifically for that person. Frith (1992) explains this as a dysfunction of ToM. According to him the mechanism for enabling metarepresentation fails and the primary representation becomes detached from the patient's knowledge of others.

There is still much debate as to how an impaired ToM in schizophrenia is associated with other aspects of cognition and, indeed, whether theory of mind is impaired or exaggerated in some types of schizophrenia. Frith (1992) has argued that a compromised theory of one's own and others' minds in schizophrenia may account for: (1) disorders of 'willed action' (e.g. negative and disorganized symptoms); (2) disorders of self-monitoring (e.g. delusions of alien control and hallucinations commenting on one's thoughts or other 'passivity' symptoms); and (3) disorders of monitoring other people's thoughts and intentions, including delusions of reference and persecution.

In relation to theory of mind, a substantial number of studies have been carried out to test the theoretical model put forward by Frith (1992). Paranoid patients have an intact ToM in the sense that they know that other people have mental states. They are, however, impaired in using contextual information, which leads them to make incorrect 'online' inferences about what these mental states are. The false beliefs found in delusional disorder are social, and involve mistaken mental state inferences such as misjudging the motivations and intentions of other people. Such mistakes are inevitable, given the nature of the Theory of Mind mechanism: beliefs concerning the mental state of others cannot always be true because beliefs cannot be checked against objective criteria – there is no direct access to other minds.

There is some evidence that patients with schizophrenia have difficulty applying strategic social rules and tactics because of their impaired ToM

(Sullivan and Allen 1999). More recently, Marianna Mazza et al. (2003) confirmed that ToM deficits in schizophrenia may be associated with impaired strategic social thinking: patients with predominantly 'positive symptoms' (hallucinations, etc.) who performed better on the ToM tasks than patients with negative symptoms (apathy, withdrawal) also had a more 'cynical' and pragmatic view of the world as measured using the Mach-IV scale, whereas patients with negative symptoms obtained lower scores in 'Machiavellianism'. Thus, impaired strategic social reasoning might reflect a deficit in appreciation of second order mental states (ibid.).

Conclusion: Schizophrenia, Religion and Creativity

I have compared religious thinking with cognition in schizophrenia and outlined some similarities and differences.[5] Specifically there are parallels in the ways in which boundaries between the self and outside world are breached. This is not to suggest that phenomenologically they are the 'same' experience or that they are associated with the same affective response: just that they depend on similar cognitive strategies involving agency and ToM. In religious cognition ToM is intact whereas in extreme religious experience there is a breakdown in the boundary between the self and outside world. In schizophrenia ToM is disordered – not only are there problems understanding other minds but there may be breaches of the self-other boundaries. And in schizophrenia, delusions of reference may be similar to the ideas held by religious believers that mundane experiences have personal significance and purpose. Both involve overattributions of intention onto the outside world.

An interesting question is whether both originate in the same evolutionary trajectory. Although, like some evolutionary explanations, we lack empirical data to establish this position, we speculate here that everyday cognition, religion and schizophrenia are on a continuum; both 'religion' and schizophrenia perhaps derive from an overattribution of agency and an overextension of ToM. There is some suggestion that schizophrenia and religion may have evolved together. Schizophrenia and schizotypy have been associated with creativity throughout recorded history (Claridge, Pryor and Watkins 1990; Nettle 2006). Psychological experiments, biographical survey studies and neurophysiology support the hypothesis that schizotypal cognition is associated with creativity and divergent thinking (Claridge, Pryor and Watkins 1990; Nettle 2006; Barrantes-Vidal 2004). Schizotypy is thought to be a disposition for schizophrenia (Claridge 1985), and schizotypic traits consist of tendencies to have abnormal perceptual or cognitive experiences.

Family studies and adoption studies indicate that schizophrenia in a family member is associated with an increased risk of schizotypy. On the other hand, it is also an indication of an increased likelihood of high creativity, leadership qualities, high musical skills and an intense interest in religion (Horrobin 1998). David Horrobin proposes that schizophrenia was probably present in the earliest stages of human development, about 150,000–100,000 years ago, around the time that there was a cultural explosion of art and religion. Furthermore schizophrenia and 'human genius' began to manifest themselves as a result of evolutionary pressures that ultimately triggered genetic changes in our brain cells, allowing us to make unexpected links with different events, an ability that significantly enhanced our intellectual abilities. Early manifestations of this creative change, according to him, include the 30,000-year-old cave paintings found in France and Spain.[6] Thus schizophrenia was the price paid for these cultural developments.[7]

A prediction of this hypothesis is that schizophrenia has evolved, and is maintained, in part as a by-product of recent positive selection and adaptive evolution in humans (Crow 1997; Horrobin 1998); creativity may be one type of 'compensatory advantage' for those carrying the genes for psychosis. Such creativity may have been associated with the potential for symbolic cognition including the possibility for the attribution of agency and ToM onto ultrahuman entities.

Simon Dein is an honorary professor at Durham University and a visiting professor at University of Glyndwr. He has written extensively on religion and health, religious experience and Jewish messianism. He teaches on an MSc on spirituality, theology and health at Durham University. He is one of the editors of the journal *Mental Health, Religion & Culture,* the chair of the Science & Faith group of the World Association of Cultural Psychiatry and he is on the executive committee of the World Psychiatric Association Transcultural Group and the Religion Group.

Acknowledgements

I am grateful to Robin Dunbar for his comments on an earlier version of this chapter.

Notes

1. While the origins of religion can only be dated with some confidence back to the practice of burial with grave goods (50,000 bce), we would argue here that its cognitive antecedents can probably be traced from the period of hominid evolutionary adaptive functioning (late Pleistocene). Archaeologists trace religion back to our earliest Sapiens progenitors (Mithen 1999; Dickson 1990). In the absence of written records or distinctive pathoanatomical features, we clearly cannot attempt to date the origin of schizophrenia or other psychoses to before the earliest medical texts (c 1000 bce), but we are proposing that it is the antecedents of both religion and schizophrenia that are significant in terms of natural selection (whilst allowing for later selection from biological advantages or disadvantages of institutions stemming from religion or schizophrenia; see note 2).
2. Once established, religious systems are, we argue, likely to be biologically advantageous (through group cohesion and mobilization) and thus persist. In other words, a social pattern is biologically advantageous, as in Fox's (1980) theory of culturally mandated incest avoidance having biological advantages through its social consequences (exchange of sisters and hence the formation of alliances).

   The origin of social formations among humans is still poorly understood. Various theories have been proposed, including kin selection in which cooperation is genetically rewarded by favouring kin (Hamilton 1964), reciprocal altruism in which reciprocal acts are returned later (Trivers 1971), indirect reciprocity in which one's reputation for cooperation is rewarded indirectly through the favour of third party observers (Nowak and Sigmund 1998), and costly signalling in which generosity signifies high fitness to mates or allies (Zahavi 1995). While these theories may explain cooperation among nonhumans, among humans cooperation still occurs even when kin selection, reciprocal altruism, indirect reciprocity and costly signalling are not immediately apparent (Gintis 2003).

   Punishment may function to ensure cohesion and cooperation (Andreoni et al. 2003). Recently, Johnson and Kruger (2004) have argued that ancestral cooperation was promoted because norm violations were deterred by the threat of supernatural punishment. There is ethnographic data to support this contention; religious believers alter their behaviours to avoid supernatural retribution (Boyer 2001). Gods, dead ancestors and other ultrahuman entities are commonly held to bring about misfortune. In contemporary evangelical Christianity and Islam misfortune is attributed to sin; only those who act contrary to God's will are punished. In many traditional societies the emphasis is on other-worldly punishment for the violation of religious norms. The question however is why gods would not want us to engage in such behaviours. When we posit that gods have minds and know the thoughts of humankind this is generally in the context of moral behaviour – gods appear to have little interest in other aspects of our life. According to the work on just-world beliefs, we operate under the assumption that others will 'get what they deserve', especially when they have little control over negative events (Lerner 1980). Bering argues that it is easy to understand how some behaviour in the moral domain is connected to an uncontrollable and unrelated life event; our innate cognitive tendency to search for reason and intention in life events predisposes us to see gods as agents who are deeply involved in our life events. Thus negative events are easily attributed to divine agency.
3. While the cognitive science of a religion approach convincingly accounts for religious cognition, we would argue that this approach to date has not developed a coherent theory of religious emotion and religious experience – a composite whole involving cognitive, affective and behavioural components. Some recent research is promising in this respect (Boyer and Lienard 2006; Whitehouse 2004).

4. See Napier, this volume.
5. In this chapter, we have restricted ourselves to considering schizophrenia rather than psychosis in general, but similar arguments could be applied to manic-depressive illness, though here alienation of agency is likely to be less profound.
6. In a previous paper (Littlewood and Dein 2013) we use the term 'protoschizophrenia' to refer to this phenomenon, arguing that our contemporary form of schizophrenia only emerged in the Common Era.
7. These mutations involved changes in lipid metabolism. Initially, they predisposed protohominids to schizotypy. It was only later (around 100,000 years ago) that mutations to the phospholipase A2 cycle gave us the potential for: schizophrenia, bipolar disorder and psychopathy. Frank psychosis was avoided by the ingestion of a fatty acid–rich water diet in riverine areas. The Industrial Revolution, with its reduction in the range and amounts of essential fatty acids, led perhaps to an explosion of psychosis in the modern world.

References

Andreoni, J., W. Harbaugh and L. Vesterlund. 2003. 'The Carrot or the Stick: Rewards, Punishments, and Cooperation.' *American Economic Review* 93: 893–902.

Atran, S. 2002. *In Gods We Trust: The Evolutionary Landscape of Religion*. Oxford: Oxford University Press.

Barrantes-Vidal, N. 2004. 'Creativity and Madness Revisited from Current Psychological Perspectives.' *Journal of Consciousness Studies* 11: 58–78.

Barrett, J. 2000. 'Exploring the Natural Foundations of Religion.' *Trends in Cognitive Sciences* 4: 29–34.

Barrett, R. 2004. 'Kurt Schneider in Borneo: Do First Rank Symptoms Apply to the Iban?' In J. Jenkins and R. Barrett Schizophrenia (eds), *Culture and Subjectivity: The Edge of Experience.* Cambridge: Cambridge University Press, pp. 87–109.

Bartocci, G., and S. Dein. 2005. 'Detachment: Gateway to the World of Spirituality.' *Transcultural Psychiatry* 42(4): 545–69.

Beit Hallahmi, B., and M. Argyle. 1997. *The Psychology of Religious Behaviour, Belief and Experience.* London: Routledge

Bentall, R., and S. Kanay. 1989. 'Contents for Specific Processing and Persecutory Delusions: An Investigation Using the Emotional Stroop Test.' *British Journal of Medical Psychology* 62: 355–64.

Bering, J. M., and D. D. P. Johnson. 2005. '"O Lord . . . You Perceive My Thoughts from Afar": Recursiveness and the Evolution of Supernatural Agency.' *Journal of Cognition and Culture* 5: 118–42.

Boyer, P. 1994. *The Naturalness of Religious Ideas: A Cognitive Theory of Religion.* Berkeley: University of California Press.

———. 2001. *Religion Explained.* London: Heineman.

Boyer, P., and P. Lienard. 2006. 'Why Ritualized Behavior? Precaution Systems and Action-parsing in Developmental, Pathological, and Cultural Rituals.' *Brain and Behavioral Sciences* 29: 595–613.

Brewerton, T. 1994. 'Hyper-religiosity in Psychotic Disorders.' *Journal of Nervous and Mental Diseases* 182: 302–4.
Brune, M. 2005. 'Theory of Mind in Schizophrenia: Review of the Literature.' *Schizophrenia Bulletin* 31: 21–42.
Claridge, G. S. (1945) Origins of Mental Illness. Oxford: Blached.
Claridge, G., R. Pryor and G. Watkins. 1990. *Sounds from the Bell Jar: Ten Psychotic Authors*. London: The Macmillan Press Ltd.
Crow, T. J. 1997. 'Is Schizophrenia the Price that *Homo Sapiens* Pays for Language?' *Schizophrenia Research* 28: 127–41.
Dawkins, R. 2008. *The God Delusion*. Boston: Houghton Mifflin.
Dein, S., and R. Littlewood. 2007. 'The Voice of God.' *Anthropology and Medicine* 14: 213–28.
Dickson, D. B. 1990. *The Dawn of Belief: Religion in the Upper Paleolithic* of *Southwestern Europe*. Tuscon: University of Arizona Press.
Dunbar, R. 2005. *The Human Story*. London: Faber and Faber.
Erlenmeyr-Kimling, L., and W. Paradowski. 1968. 'Selection of Schizophrenia.' *American Naturalist* 100: 651–65.
Fabrega, H. 1982. 'Culture and Psychiatric Illness: Biomedical and Ethnomedical Aspects.' In A. Marsella and G. White (eds), *Cultural Conceptions of Mental Health and Therapy*. Dordrecht: Reidel, pp. 39–68.
Feuerbach, L. 1967. *Lectures on the Essence of Religion*. New York: Harper and Row.
Fish, F. 1967. *Clinical Psychopathology*. London: Wright.
Fox, R. 1980. *The Red Lamp of Incest*. New York: Dutton.
Frith, C. 1992. *The Cognitive Neuropsychology of Schizophrenia*. Hove, East Sussex: Lawrence Erlbaum.
Gergely, G., and G. Csibra, G. 2003. 'Teleological Reasoning in Infancy: The Naive Theory of Rational Action.' *Trends in Cognitive Sciences* 7: 287–92.
Gintis, H. 2003. 'Solving the Puzzle of Prosociality.' *Rationality and Society* 15: 155–87.
Gould, S. 1997. 'The Exaptive Excellence of Spandrels as a Term and Prototype.' *Proceedings of the National Academy of Sciences USA* 94: 10750–10755.
Guthrie, S. 1993. *Faces in the Clouds: A New Theory of Religion*. Oxford: Oxford University Press.
Heider, F., and M. Simmel. 1944. 'An Experimental Study of Apparent Behaviour.' *American Journal of Psychology* 57: 243–59.
Horrobin, D. F. 1998. 'Schizophrenia: The Illness that Made Us Human.' *Medical Hypotheses* 50: 269–88.
James, W. 1958. *The Varieties of Religious Experience*. New York: Simon & Schuster.
Janet, P. 1937. 'Les Troubles de la Personnalite Sociale.' *Annales Medico-Psychologiques* 11: 149–200.
Johnson, D. D. P., and O. Kruger. 2004. 'The Good of Wrath: Supernatural Punishment and the Evolution of Cooperation.' *Political Theology* 5(2): 159–76.
Karlsson, J. L. 1984. 'Creative Intelligence in Relatives of Mental Patients.' *Hereditas* 100: 83–86.
Kurzban, R., J. Tooby and L. Cosmides. 2001. 'Can Race be Erased? Coalitional Computation and Social Categorization.' *Proceedings of the National Academy of Sciences* 98(26): 15387–15392.
La Barre, W. 1970. *The Ghost Dance*. London: Allen and Unwin.

Lerner, M. 1980. *The Belief in a Just World*. New York: Plenum.
Lienhardt, G. 1962. *Divinity and Experience: The Religion of the Dinka*. Oxford: Clarendon Press.
Littlewood, R. 1993. *Pathology and Identity: The Work of Mother Earth in Trinidad*. Cambridge: Cambridge University Press.
———. 1996. 'Psychopathology, embodiment and religious innovation: An historical instance.' In D. Bhugra (ed.), *Religion and Mental Illness*. London: Routledge, pp. 178–97.
Littlewood, R., and S. Dein. 2013. 'Did Christianity Lead to Schizophrenia? Psychosis, Psychology and Self Reference.' *Transcultural Psychiatry* 50(3): 397–420.
Luhrmann, T. M. 2005. 'The Art of Hearing God: Absorption, Dissociation and Contemporary American Spirituality.' *Spiritus: A Journal of Christian Spirituality* 5: 133–57.
Mazza, D., et al. 2001. 'Selective Impairments of Theory of Mind in People with Schizophrenia.' *Schizophrenia Research* 47: 299–308.
Mazza, M., et al. 2003. 'Machiavellianism and Theory of Mind in People Affected by Schizophrenia.' *Brain and Cognition* 51: 262–69.
Michotte, A. 1963. *The Perception of Causality* (translated by T. R. Miles and E. Miles). New York: Basic Books.
Mithen, S. 1999, 'Symbolism and the Supernatural.' In R. I. Dunbar, C. Knight and C. Power (eds), *The Evolution of Culture*. New Brunswick, NJ: Rutgers University Press.
Moslowski, J., D. Jansenvanrensberg and N. Mthok. 1998. 'A Poly-Diagnostic Approach to the Differences in Symptoms of Schizophrenia in Different Cultural and Ethnic Populations.' *Acta Psychiatrica Scandinavia* 98: 41–46.
Mohr, S., et al. 2006. 'Towards the Integration of Spirituality and Religiousness into the Psychosocial Dimension of Schizophrenia.' *American Journal of Psychiatry* 163: 1952–59.
Nettle, D. 2006. 'Schizotypy and Mental Health Amongst Poets, Visual Artists, and Mathematicians.' *J. Res. Pers* 40: 876–90.
Nowak, M. A., and K. Sigmund. 1998. 'Evolution of Indirect Reciprocity by Image Scoring.' *Nature* 393: 573–77.
Otto, R. 1958. *The Idea of the Holy*. New York: Oxford University Press.
Pettazzoni, R. 1955. 'On the Attributes of God.' *Numen* 2: 1–27.
Peters, E., et al. 1999. 'Delusional Ideation in Religious and Psychotic Populations.' *British Journal of Clinical Psychology* 38: 83–96.
Pickering, W. 1994. *Durkheim on Religion*. Atlanta, GA: Scholars Press.
Polimeni, J. and J. Reiss. 2002. 'How Shamanism and Group Selection May Reveal the Origins of Schizophrenia'. *Medical Hypothesis*, 58:244-8
———. 2003. 'Evolutionary Approaches on Schizophrenia.' *Canadian Journal of Psychiatry* 48: 34–39.
Schneider, K. 1955. *Klinische Psychopathologie*. Studdgart: Thieme Verlag.
Schumaker, J. 1995. *The Corruption of Reality: A Unified Theory of Religion, Hypnosis and Psychopathology*. New York: Prometheus Books.
Sell, A. N., et al. (2003). *Encyclopedia of Cognitive Science*. London: Macmillan, pp. 47–53.

Siddle, R., et al. 2002. 'Religious Delusions in Patients Admitted to Hospital with Schizophrenia.' *Social Psychiatry and Psychiatric Epidemiology* 37: 130–38.
Silverman, I., and M. Eals. 1992. 'Sex Differences in Spatial Abilities: Evolutionary Theory and Data.' In J. Barkow, L. Cosmides and J. Tooby (eds), *The Adapted Mind: Evolutionary Psychology and the Generation of Culture*. Oxford: Oxford University Press, pp. 533–49.
Simon, B. 1978. *Mind and Madness in Ancient Greece*. Ithaca, NY: Cornell University Press.
Sosis, R., and C. Alcorta. 2003. 'Signalling, Solidarity and the Sacred: the Evolution of Religious Behaviour.' *Evolutionary Anthropology* 12: 264–74.
Stevens, A., and J. Price. 2000. *Evolutionary Psychiatry: A New Beginning*. London: Routledge.
Thalbourne, M. A., and P. S. Delin. 1994. 'A Common Thread Underlying Belief in the Paranormal, Mystical Experience and Psychopathology.' *Journal of Parapsychology* 58: 3–38.
Tracy, M. 2005. 'Postpartum Depression: An Evolutionary Perspective.' *Nebaska Anthropologist* 20.
Trivers, R. L. 1971. 'The Evolution of Reciprocal Altruism.' *Quarterly Review of Biology* 46: 35–57.
Ulman, L., and L. Krasler. 1969. *A Psychological Approach to Abnormal Behaviour*. Oxford: Prentice Hall.
Verplaetse, J., S. Venneste and J. Braeckman. 2007. 'You Can Judge and Book by its Cover: The Sequel. A Kernel of Truth in Predictive Cheating Detection.' *Evolution and Human Behaviour* 28: 260–71.
Whitehouse, H. 2004. *Modes of Religiosity: A Cognitive Theory of Religious Transmission*. Walnut Creek, CA: AltaMira Press.
Zahavi, A. 1995. 'Altruism as a Handicap: The Limitations of Kin Selection and Reciprocity.' *Journal of Avian Biology* 26: 1–3.

Index

D

E

F

G

H

I

J